Intelligent Materials for Controlled Release

ACS SYMPOSIUM SERIES **728**

Intelligent Materials for Controlled Release

Steven M. Dinh, EDITOR
Lavipharm Inc.

John D. DeNuzzio, EDITOR
Becton Dickinson Research Center

Ann R. Comfort, EDITOR
Novartis Pharmaceuticals Corporation

American Chemical Society, Washington, DC

Library of Congress Cataloging-in-Publication Data

Intelligent materials for controlled release / Steven M. Dinh, John D. DeNuzzio, Ann R. Comfort, editors.

p. cm—(ACS symposium series : 728)

Includes bibliographical references and index.

ISBN 0–8412–3595–3

1. Drugs—Controlled release. 2. Polymeric drug delivery systems. 3. Drug carriers (Pharmacy)

I. Dinh, Steven M. II. DeNuzzio, John D., 1959— III. Comfort, Ann R., 1960— . IV. Series.

RS201.C64I56 1999   99–12879
615.7—dc21   CIP

The paper used in this publication meets the minimum requirements of American National Standard for Information Sciences—Permanence of Paper for Printer Library Materials, ANSI Z39.48-94 1984.

PRINTED IN THE UNITED STATES OF AMERICA

Foreword

THE ACS SYMPOSIUM SERIES was first published in 1974 to provide a mechanism for publishing symposia quickly in book form. The purpose of the series is to publish timely, comprehensive books developed from ACS sponsored symposia based on current scientific research. Occasionally, books are developed from symposia sponsored by other organizations when the topic is of keen interest to the chemistry audience.

Before agreeing to publish a book, the proposed table of contents is reviewed for appropriate and comprehensive coverage and for interest to the audience. Some papers may be excluded in order to better focus the book; others may be added to provide comprehensiveness. When appropriate, overview or introductory chapters are added. Drafts of chapters are peer-reviewed prior to final acceptance or rejection, and manuscripts are prepared in camera-ready format.

As a rule, only original research papers and original review papers are included in the volumes. Verbatim reproductions of previously published papers are not accepted.

ACS BOOKS DEPARTMENT

Contents

Preface

Modern drug delivery and pharmaceutical systems are incorporating new levels of intelligence in their design and performance. New materials are designed to respond to their environment; new miniaturized systems perform multiple tasks to analyze samples and deliver drugs; and new drug formulations enable targeted therapy to specific organs. This book spans work in many diverse areas of biomedical and pharmaceutical sciences which demonstrate the development and use of intelligent materials and systems.

This book is the result of a symposium entitled "Intelligent Materials and Novel Concepts for Controlled Release Technologies," held during the ACS National Meeting in San Francisco, California in April 1997. The Symposium brought together academic and industrial scientists who had developed novel drug formulations, techniques to enhance delivery, and new systems and procedures to analyze biomolecules. We hope that bringing together key researchers in these diverse fields and initiating direct new dialogues between them will lead to next-generation intelligent systems for drug delivery and biomedical applications.

STEVEN M. DINH
Lavipharm
69 Princeton-Hightstown Road
Hightstown, NJ 08520

JOHN D. DENUZZIO
Becton Dickinson Research Center
P.O. Box 12016
Davis Drive
Research Triangle Park, NC 27709

ANN R. COMFORT
Novartis Pharmaceuticals Corporation
59 Route 10
East Hanover, NJ 07936

Chapter 1

Cogelation of Hydrolyzable Cross-Linkers and Poly(ethylene oxide) Dimethacrylate and Their Use as Controlled Release Vehicles

Jennifer Elisseeff[1], Winnette McIntosh[1], Kristi Anseth[2], and Robert Langer[1]

[1]Harvard–MIT Division of Health Sciences and Technology
and Department of Chemical Engineering, Massachusetts Institute of Technology,
Cambridge, MA 02139
[2]Department of Chemical Engineering, University of Colorado, Boulder, CO 80301

Small molecular weight crosslinking agents containing hydrolyzable bonds were photopolymerized with poly(ethylene oxide) in order to decrease the pore size of the gels. The resulting cogels decreased in equilibrium swelling volume (pore size) as the concentration of crosslinker increased. The initial release profile of the model protein albumin showed a decreased burst in the presence of small molecular weight crosslinkers in the photopolymerized hydrogels.

Hydrogels are crosslinked macromolecule networks swollen in water or biological fluids. Among other applications, hydrogels have been utilized for drug delivery for many years. The release of molecularly dissolved or dispersed drugs at high concentrations in polymer matrices has been examined in swelling-controlled release systems [1] . Previously, hydrogels were synthesized, dried and reswollen in order to load the drug of interest [2] . Prefabrication of the polymer was necessary since the gels were typically formed by thermal crosslinking polymerizations or radiation crosslinking. Most drug molecules were unstable in these relatively harsh processing conditions. Recently, new advances and methods in hydrogel synthesis have allowed *in situ* encapsulation of drugs during hydrogel synthesis [3] . *In situ* encapsulation simplifies fabrication methods by eliminating the post-gelation drug loading step and also provides uniform drug distributions on the time scale of minutes, rather than hours or days. Photopolymerization is one such technique employed to fabricate a hydrogel and is the focus of this contribution.

Photopolymerization is a method frequently used in the dental community to form dental sealents [4] . Currently, photocurable dimethacrylate composites are cured *in situ* in the dental carie using blue light and visible light initiators. These resins react to form highly crosslinked networks that are dimensionally stable and mechanically strong and provide aesthetic alternatives to mercury amalgam fillings. The kinetics and biocompatiblity of photopolymerizations have been examined thoroughly for this application [4, 5] . More recently, new biomedical applications of *in vivo* photopolymerization have been examined [6] . The release of poly(ethylene oxide) based hydrogels has been examined in preventing post surgical adhesions and in preventing restenosis in artery walls after angioplasty [7,

8, 9] . Finally, Grubbs has developed and used photocurable polymers to replace endocapsular lenses that opacify as the result of cataract development [10] . New methods of photopolymerizing polymers through skin and other tissues using both UVA and visible light are currently being examined [11] . The present work describes the use of small molecular weight crosslinking agents which can alter the swelling and release characteristics of photopolymerized poly(ethylene oxide) hydrogels.

Experimental

Synthesis of Succinic Dimethacrylate (SADM). Succinic acid was dissolved in anhydrous dimethyl sulfoxide (DMSO, Aldrich) to which an excess of methacrylic anhydride (Aldrich) was added. The reaction mixture was purged with argon and heated to 40°C for 24 h. The reaction mixture was cooled to room temperature and precipitated by adding to a 10 X excess of ether. The precipitate was filtered and dried under vacuum. Infrared spectroscopy (IR, Nicolet, Inc.) was utilized to analyze the product in a KBr pellet.

Hydrogel Synthesis. Poly(ethylene oxide) dimethacrylate (PEODM), MW 1000 and 3400, were purchased from Polysciences and Shearwater respectively. N,N'-diallyltartardiamide (DAT) and N,N'-(1,2-dihydroxyethylene) bisacrylamide (DEB) were purchased from Polysciences. Varying percentages of DAT, DEB or SADM were dissolved in water with PEODM to form a 50/50% w/v solution using on the order of 100 mg PEODM. The polymer solution was subsequently exposed to UV radiation (EFOS Ultracure) in the presence of a radical photoinitiator (2,2-dimethoxy-2-phenyl acetophenone or 2-hydroxycylcohexyl ketone) for 1 minute. Macromer solutions containing more than 40% SADM, DEB or DAT were heated on a hot plate before photopolymerization to dissolve the crosslinkers.

Swelling and Elemental Analysis. The cogels were swollen in 3 mL phosphate buffered saline (PBS). Swollen weights increased and stabilized after 2 d. The equilibrium swelling volume, Q, was calculated using the 2d swelling weight for the DEB and DAT cogels. Values shown are an average of four samples with standard deviations. One hour swelling weights were taken for the SADM cogel calculations due to the fast degradation of the SADM anhydride bond. Elemental analysis (C,H,N) on the DEB, DAT cogels was performed by Quantitative Technologies Inc (average of three samples with standard deviation). Gels were dried and ground for elemental analysis.

Release Studies. Bovine serum albumin (BSA, Sigma) was added to the 50/50% w/v macromer solution of MW 1000 PEODM and a crosslinking agent (SADM, DAT or DEB), vortexed and photopolymerized 1 minute in scintillation vials to which 3 mL PBS was added. A 5% (w/w) loading dose of albumin was encapsulated in the hydrogels. The gels were incubated at 37°C (static). At various time points the PBS was removed and frozen while 3 mL fresh PBS was added. Albumin was quantitated using a micro BCA assay (Pierce). Rhodamine was encapsulated in a similar fashion and release was observed by flourimetry.

Results and Discussion

Synthesis of Hydrogels. The molecular weight of a poly(ethylene oxide) dimethacrylate macromer controls many properties of the resulting macromer solution. For example, high molecular weight PEODM forms a viscous macromer solution amenable to injection and *in vivo* polymerization whereas lower molecular weight PEODM macromer solutions are less viscous and may leak into surrounding tissue to a large degree before polymerization. Low molecular weight PEODM forms rigid gels which in addition have been shown to calcify faster than higher molecular weight counterparts when placed subcutaneously in rats [12]. For these reasons, the addition of small amounts of crosslinkers which can slow initial burst release, yet do not change the properties of high molecular weight macromer solutions to a large extent are desired. The crosslinkers chosen also have potentially hydrolyzable bonds to aid hydrogel degradation.

The substitution of an acid group with methacrylic anhydride has been previously described as a method to incorporate methacrylate groups on aspartic acid residues and anhydrides [13]. Succinic acid was substituted employing methacrylic anhydride in a similar manner. Infrared spectroscopy was used to examine the substitution of the succinic acid carboxylic acid groups. Figure 1 compares the IR of succinic acid before and after reacting with methacrylic anhydride and shows the disappearance of the wide acid peak centered at $3100cm^{-1}$. Complete substitution was not achieved due to the difficulty in removing residual water, which intereferes with anhydride formation, from succinic acid.

Succinic acid was chosen as a crosslinking agent due to its water solubility and the extremely labile anhydride bond which is formed during the attachment of the methacrylate groups. Two other small molecular weight crosslinking agents, DAT and DEB, were chosen to copolymerize with PEODM. They are soluble in water and in the PEODM macromer solution at low concentrations and contain bonds which are hydrolyzable under mild conditions. Figure 2 illustrates the structures of the three crosslinking agents SADM, DAT and DEB along with PEODM.

The equilibrium swelling volume, Q, is a characteristic of a hydrogel which depends on the ability of the network to absorb water. The equilibrium swelling is defined as the volume of the swollen network divided by the volume of the dry network. The amount of water the hydrogel is able to absorb is dependent on the pore size of the network. The average molecular weight between crosslinks can be estimated using Q and models derived by Flory and further modified by Brannon et al [14, 15]. For example, when small molecular weight crosslinking agents such as SADM, DEB and DAT are added to PEODM, the molecular weight between crosslinks decreases and the swelling should also decrease. As more of the crosslinking agent is added, the swelling decrease even further. The equilibrium swelling volume was determined for cogels made from SADM, DAT, DEB and PEODM in order to observe if the crosslinker was being incorporated into the network and effectively decreased the pore size. Figure 3 portrays the decrease in Q with increasing levels of SADM for PEODM of molecular weights 3400 and 1000. The change in equilibrium swelling of PEO cogels with DAT and DEB is shown in Figure 4. The addition of one mole of DEB per mole of PEO decreased the swelling 42%. Ten moles of DEB per mole of PEO decreased the swelling 81%. The larger percentages of DAT (ten moles per mole PEO) did not decrease to as a large a degree due to the lack of water solubility at high concentrations. Note that the equilibrium swelling of PEODM with no crosslinker present was 6.5, much larger than the value of 2.8 shown in Figure 3. The larger value was obtained after 2 days of swelling stabilization whereas in the case of the SADM hydrogels, the swollen weight was obtained after one hour in order to compare to gels containing SADM.

Although the decrease in the equilibrium swelling volume implies incorporation of the crosslinking agents into the PEODM hydrogels, elemental analysis was performed in order to further quantify incorporation. Also, DAT and DEB have polymerizing endgroups which have different reactivities compared to the PEODM methacrylate groups potentially decreasing their incorporation in the polymer network. Nitrogen is present only in the crosslinkers DEB and DAT and was therefore used to compare their concentrations relative to PEO in gels. Acrylamide groups have a higher reactivity than the allyl groups. This difference in reactivity was observed in the cogels by a higher incorporation of DEB compared to DAT in

4

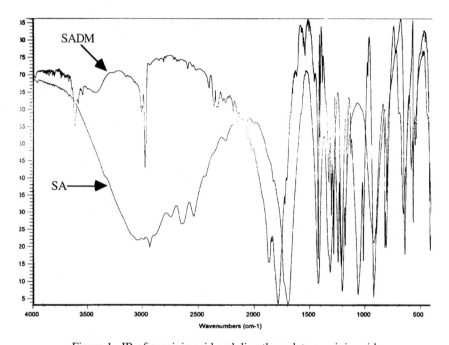

Figure 1. IR of succinic acid and dimethacrylate succinic acid.

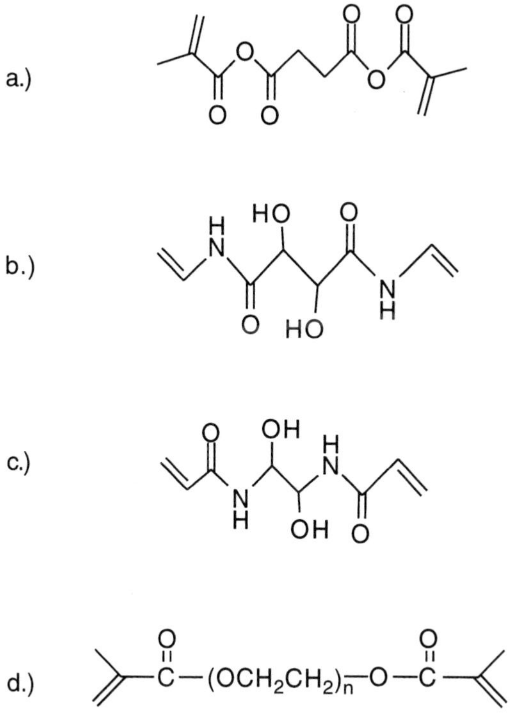

Figure 2. Monomer structures of a.) succinic acid dimethacrylate (SADM) b.) N,N'-diallyltartardiamide (DAT) c.) N,N'-(1,2-dihydroxyethylene) bisacrylamide (DEB) and d.) poly(ethylene oxide) dimethacrylate (PEODM).

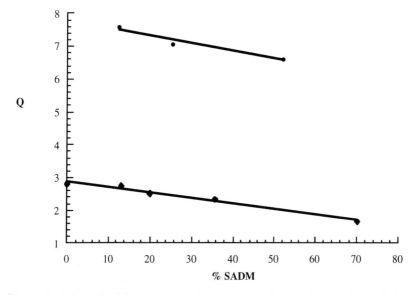

Figure 3. Effect of SADM concentration on equilibrium swelling volume, Q, in cogels of PEODM, MW 3400 and 1000.

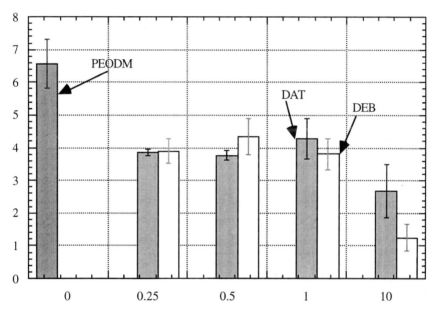

Figure 4. Effect of DAT and DEB on equilibrium swelling volume, Q, in cogels of PEODM MW 1000.

the cogels (Figure 5). At low concentrations of crosslinking agent, the experimental values of crosslinker incorporation determined by elemental analysis were closer to the calculated macromer concentration values of crosslinker in the gels. The lower endgroup reactivities and insolubility of the crosslinkers, DAT and DEB, decreased their incorporation in cogels at higher crosslinker concentrations.

Release of Hydrogels. Bovine serum albumin (BSA) was used as a model protein to examine the effects of the addition of small molecular weight crosslinkers to PEODM networks. The early release profiles of hydrogels with 0, 17 and 43% SADM are depicted in Figure 6. As the concentration of SADM in the gels increases, the amount of albumin released decreases. For the 43% SADM gel, approximately 20% less albumin is released at early time points. The gels were allowed to release for up to 100 days (not shown). At these later times, the amount of albumin released (approximately 80%) was equal for all of the SADM gels. Rhodamine (MW 457) was encapsulated to model the release of a small molecule. Eighty percent of encapsulated rhodamine was released in one week in 17% SADM gels.

The addition of approximately one mole DEB per mole of PEODM decreased initial (8 day) release by 20-30% (Figure 7). DAT showed a smaller decrease in initial release, 10-20% less albumin was released in cogels of DAT (Figure 8). At later timepoints (40 days) this difference in albumin released remained, unlike in the case of SADM gels.

Conclusion

The addition of small crosslinking agents such as succinic dimethacrylate, N,N'-diallyltartardiamide and N,N'-(1,2-dihydroxyethylene)bisacrylamide to poly(ethylene oxide) dimethacrylate photopolymerized hydrogels, changes the swelling properties of the resulting polymer networks. The initial burst release of encapsulated bovine serum albumin decreases when these small crosslinking molecules are added to PEODM gels.

These crosslinking agents allow the pore size and release characteristics of a hydrogel to be altered while using only one molecular weight PEODM macromer. This may provide useful in clinical applications where certain molecular weight PEO chains must be used to have a viscous macromer solution and a resulting gel which will not undergo calcification.

Acknowledgment

This work is funded by Advanced Tissue Sciences and NSF Grant No. 9202311.

Literature Cited

1. Colombo, P., R. Bettini, G. Massimo, P. Catellani, P. Santi and N. Peppas. "Drug Diffusion front movement is important in drug release control from swellable matrix tablets." *J. of Pharm. Sci,.* **84**: 991-97, 1995.
2. Bell, C. L. and N. Peppas. "Biomedical Membranes from Hydrogels and Interpolymer Complexes." *Advances in Polymer Science.* **122**: 128-139, 1995.

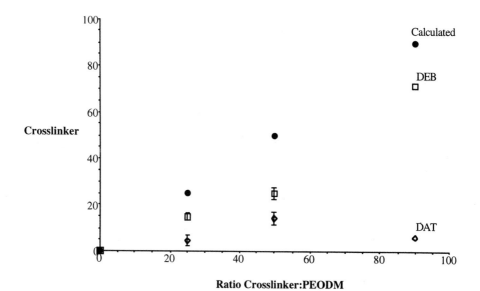

Figure 5. Amount of DAT and DEB incorporated in gels determined by elemental analysis.

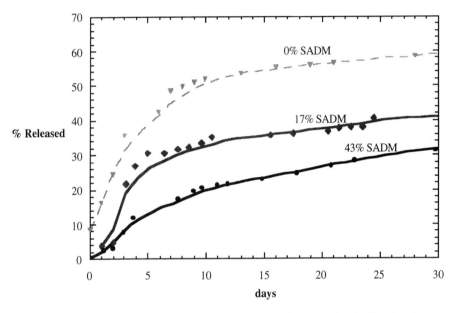

Figure 6. Percent bovine serum albumin released from cogels of SADM and PEODM over the first 30 days.

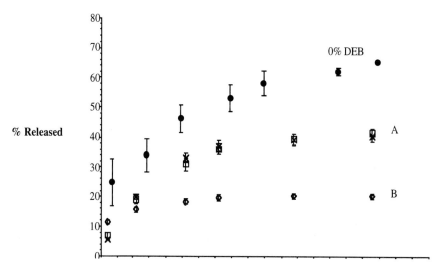

Figure 7. Percent bovine serum albumin released from cogels of DEB and PEODM for initial 200 hours (8.3 days).

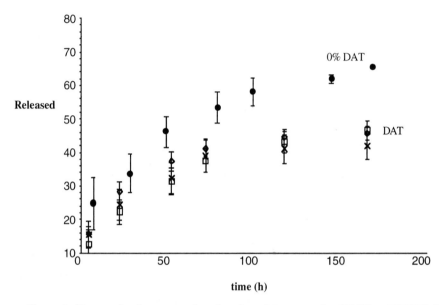

Figure 8. Percent bovine serum albumin released from cogels of DAT and PEODM for initial 200 hours (8.3 days).

3. Hill-West, J., R. Dunn and J. Hubbell. "Local release of fibrinolytic agents for adhesion prevention." *J Surg Res.* **59**(6): 759-63, 1995.

4. Anseth, K., S. Newman and C. Bowman. "Polymeric Dental Composites: properties and reaction behavior of multimethacrylate dental restorations." *Adv. Polym. Sci.* **122**: 177-218, 1995.

5. Venhoven, B. A. M., A. J. de Gee and C. L. Davidson. "Light initiation of dental resins: dynamics and polymerization." *Biomaterials.* (17): 2313-2318, 1996.

6. Sawhney, A., C. Pathak and J. Hubbell. "Bioerodible Hydrogels Based on Photopolymerized Poly(ethylene glycol)-*co*-poly(α-hydroxy acid) Diacrylate Macromers." *Macromolecules.* **26**: 581-587, 1993.

7. Sawhney, A., F. Lyman, F. Yao, M. Levine and P. Jarrett. *A Novel in situ Formed Hydrogel for use as a Surgical Sealent or Barrier. 23rd International Symposium of Controlled Release of Bioactive Materials.* (Controlled Release Societym Inc., Kyoto, Japan, 1996).

8. Hill-West, J., S. Chowdhury, M. Slepian and J. Hubbell. "Inhibition of thrombosis and intimal thickening by in situ photopolymerization of thin hydrogel barriers." *Proc Natl Acad Sci USA.* **91**(13): 5967-71, 1994.

9. Hill-West, J., S. Chowdhury, A. Sawhney, C. Pathak, R. Dunn and J. Hubbell. "Prevention of postoperative adhesions in the rat by in situ photopolymerization of bioresorbable hydrogel barriers." *Obstet Gynecol.* **83**(1): 59-64, 1994.

10. Grubbs, R., R. Coots and S. Pine. *Synthetic polymers for endocapsular lens replacement.* (U.S. 4,919,151, April 1990).

11. Elisseeff, J., D. Sims, K. Anseth W. McIntosh, M. Randolph, M. Yaremchuk and R. Langer. "Transdermal Photopolymerizations." *in preparation.* :

12. Hossainy, S. F. A. and J. A. Hubbell. "Molecular weight dependence of calcification of polyethylene glycol hydrogels." *Biomaterials.* **15**(11): 921-925, 1994.

13. Elisseeff, J., K. Anseth, R. Langer and J. Hrkach. "Synthesis of Photocrosslinked Polymers Based on Poly(L-lactic acid-*co*-aspartic acid)." *Macromolecules.* **30**(7): 2182-2184, 1996.

14. Flory, P. J. "Principles of Polymer Chemistry." 1957 George Banta Company. Menasha, WI.

15. Brannon-Peppas, L. "Preparation and Characterization of Crosslinked Hydrophilic Networks." *Superabsorbant Polymers: Science and Technology.* 1994 ACS. Washington, DC.

Chapter 2

Temperature-Sensitive Polymer System Constructed with Sodium Alginate and Poly(*N,N*-dimethylaminoethyl methacrylate-*co*-acrylamide)

S. H. Yuk[1], S. H. Cho[1,2], H. B. Lee[1], and M. S. Jhon[2]

[1]Advanced Materials Division, Korea Research Institute of Chemical Technology,
100 Jang Dong, Yusung, Taejeon, Korea 305–600
[2]Department of Chemisty and Center for Molecular Science,
Korea Advanced Institute of Science and Technology,
23 Guseong Dong, Yusung Ku, Taejeon, Korea 305–701

Novel semi-interpenetrating polymer networks (IPNs) constructed with sodium alginate (SA) and poly (N,N-dimethylaminoethyl methacrylate (DMAEMA)-co-acrylamide (AAm)) have been prepared in an attempt to provide temperature-sensitive polymer systems. The role of SA is to provide a crosslinked network and that of poly DMAEMA-co-AAm is to provide temperature-sensitivity to the polymer network. Firstly, poly DMAEMA-co-AAm were prepared to demonstrate a temperature-induced phase transition. Poly DMAEMA has a lower critical solution temperature (LCST) around 50 °C in water. With copolymerization of DMAEMA with AAm, the LCST shift to the lower temperature was observed, probably due to the formation of hydrogen bonds between amide and N,N-dimethylamino groups. FT-IR studies clearly showed the formation of hydrogen bonds which protect N,N-dimethylamino groups from exposure to water and result in a hydrophobic contribution to the LCST. Two types of polymer gels were prepared. The one is a crosslinked gel composed of poly DMAEMA-co-AAm. The other is a composite gel composed of SA and poly DMAEMA-co-AAm which have semi-IPNs structure. Although both polymer systems exhibited the temperature-sensitivity, more rapid and significant temperature-sensitivity was observed in the case of composite gels, which exhibited significant temperature-sensitive release pattern in response to pulsatile temperature change.

Phase transitions in polymer gels have attracted much attention because of their scientific interest and technological significance. Especially, temperature-induced phase transitions in polymer gel were intensively studied by numerous scientists. Hoffman et al., Kim et al., and Okano et al. have studied aqueous swelling and drug release from poly (N-isopropylacrylamide) gels (1-5). New types of materials with transitions resulting from both polymer-water and polymer-polymer interactions have been reported. Tanaka et al. studied the volume phase transition in a gel driven by hydrogen bonding (6) and Okano et al. demonstrated a temperature-induced phase transition of polymer networks constructed with poly (acrylic acid) and poly (dimethylacryl amide) by hydrogen bonding (7).

The purpose of this study is to demonstrate a temperature-sensitive polymer gel with transitions resulting from both polymer-water and polymer-polymer interactions and to propose a temperature-induced transition mechanism. For this purpose, copolymers of N, N-dimethylaminoethyl methacrylate (DMAEMA) and acrylamide (AAm) were prepared and characterized as a function of the copolymer composition. Based on the solution properties of copolymers (8), polymer gels with two types of structures such as crosslinked polymer networks and semi-interpenetrating polymer networks (IPNs) were prepared and their swelling behaviors and solute release patterns from them were observed.

Experimental

Materials.   AAm monomer and sodium alginate (SA) were purchased from Junsei Chemical Co. (Tokyo, Japan). DMAEMA monomer, ammonium persulfate (APS), N, N-methylene-bisacrylamide (MBAAm), and tetramethylethylene diamine (TEMED) were purchased from Aldrich (Milwaukee, WI, USA). Hydrocortisone was purchased from Sigma Chemical Co. (St. Louis, MO, USA) DMAEMA monomer was distilled before use. Other reagents were used as received.

Copolymers.   Copolymers of DMAEMA and AAm were prepared by free radical polymerization in water at room temperature using APS as the initiator and TEMED as the accelerator. The feed compositions for copolymers are shown in Table I. The initiator and accelerator concentrations were 0.2 g/ml of APS and 240 μl /ml of TEMED, respectively. All polymers were purified by dialysis against distilled-deionized water and subsequent lyophilization.

Polymer Hydrogels.   Copolymer gels composed of DMAEMA and AAm were prepared by free radical polymerization using MBAAm as a crosslinker based on the feed composition in Table I.
Composite gels of SA and poly DMAEAM-co-AAm were prepared as follows: The known amounts of 2 wt% SA aqueous solution and 25 wt% poly DMAEMA-co-AAm aqueous solution were mixed thoroughly with (or without) hydrocortisone using a homogenizer (Janke & Kunkel GmbH, Staufen, Germany). A 5 wt% calcium chloride solution was poured gently onto the surface of the polymer solution mixture. The calcium chloride solution used was twice the volume of the polymer solution mixture. Because SA formed a gel in the presence of divalent ion, the gel network was formed from the surface with the penetration of calcium chloride into the polymer solution mixture. The polymer solution mixture was completely converted to the gel network within 1 hour. The gel network was immersed in distilled-deionized water for 3 hours to remove unreacted calcium chloride. For the determination of loading amount, 250 mg of dried gel network with hydrocortisone was dissolved in 250 ml of EDTA (ethylenediaminetetraacetic acid) solution for 2 hours with vigorous stirring. EDTA solution consisted of 1.2 g of EDTA·2Na and 10 g of EDTA·3Na in 100 ml of water. 5 ml aliquots were withdrawn and immediately filtered through a 0.2 μm membrane filter (Spectrum Medical Industries Inc., Los Angeles, CA, USA). The drug concentration was assayed at 248 nm using a UV spectrophotometer (Shimadzu Corp., Kyoto, Japan). The drug loading amount relative to dried gel network was approximately 20 wt%.
It was well known that the crosslinked SA (calcium alginate) was easily disintegrated into SA in EDTA or alkaline solution (decoupling). To control the crosslinking density, a composite gel (0.5 g of dried polymer) was treated with 100 ml of 10 wt % sodium acetate aqueous solution. If EDTA or a strong base such as NaOH was used as a decoupling agent, the complete disintegration occurred within 1 minute. It was hard to control the swelling or crosslinking density of the gel network

Table I. Feed Composition for Copolymers[a] in the Study

Code	DMAEMA		AAm		$M_w/10^5$ [b]
	g	mol%	g	mol%	
Poly DMAEMA	14.2	100	–	–	2.8
Copolymer I	11.4	80	1.46	20	3.2
Copolymer II	9.58	67	2.40	33	3.8
Copolymer III	7.15	50	3.65	50	3.5
Copolymer IV	5.72	40	4.38	60	3.0
Poly AAm	–	–	7.20	100	

[a] Polymers were synthesized in 90 mL of H_2O.

[b] Measured by laser scattering.

consistently because the experimental error caused by the variation of decoupling time was significant. Therefore, weak base such as sodium acetate was selected as a decoupling agent for slow decoupling to control the swelling or crosslinking density of the gel network consistently. The composite gel was equilibrated with distilled water to remove the unreacted sodium acetate after decoupling. To analyze the crosslinking density, the concentration of calcium ion in the polymer was assayed at 422.7 nm using Perkin Elmer 2380 Atomic Absorption Spectrometer (Perkin Elmer Corp., Norwalk, CT, USA). The sample was prepared by the dry-ash method.

All the swollen gels were cut into cylinders (thickness: 1 cm and diameter: 2 cm). All the gels were prepared just before the swelling or release experiments.

Transmittance Measurements. The phase transition was traced by monitoring the transmittance of a 500 nm light beam on a Spectronic 20 spectrophotometer (Baush & Lomb Inc., Sunland, CA, USA). The concentration of the aqueous polymer solution was 5 wt % and the temperature was raised from 15 to 70 °C in 2-degree increments every 10 minutes.

FT-IR Measurements. For Fourier transform infrared (FT-IR) measurements, thin films of polymers were cast from 0.5 wt % distilled-deionized water onto separate CaF_2 plates at room temperature. Most of the water in the films was removed by evaporation at 50 °C in a vacuum oven for 24 hours. FT-IR spectra of the dried polymer were measured on a Magna IR spectrophotometer (Nicolet Inc., Madison, WI, USA) using 64 average scans at a resolution of 4 cm[-1].

Swelling Measurements. After immersion in water at a desired temperature, the gel was removed from the water and tapped with a filter paper to remove excess water on the gel surface. The gel was repeatedly weighed and reimmersed in water at a fixed temperature until the hydrated weight reached a constant value. After equilibration at one temperature, the gel was reequilibrated at a desired temperature. The swelling, defined as the weight of water uptake per unit weight of dried gel, was calculated by measuring the weight of swollen gel until weight changes were within 1 % of the previous measurement.

Drug Release. The released amount of hydrocortisone from gel in response to pulsatile temperature change was measured by taking 1 ml of the release media (distilled-deionized water) at desired temperatures (20 and 38 °C), replacing the total release media (50 ml) with a fresh one to maintain sink conditions. The drug concentration was assayed at 248 nm using a UV spectrophotometer (Shimadzu Corp., Kyoto, Japan).

Results and Discussion

Temperature-induced Phase Transition of Poly DMAEMA-co-AAm Aqueous Solution. Prior to observing the temperature-sensitive behavior of polymer gels, temperature-induced phase transition of poly DMAEMA-co-AAm, which was the major part of temperature-sensitive polymer gel, was characterized.

The effect of AAm content on lower critical solution concentration (LCST) is shown in Figure 1. When the temperature of a poly DMAEMA aqueous solution was raised above 50 °C, the polymer precipitated from the solution. This is due to the hydrophobic interaction between N,N-dimethylaminoethyl groups above LCST. With the incorporation of AAm in the copolymer, LCST was shifted to a lower temperature.

FT-IR studies were used to characterize the intra/intermolecular interaction in the copolymer. Figure 2 shows a characteristic absorption band of poly AAm in the N-H stretching region. According to Coleman et al. (9-10), the spectrum of poly AAm

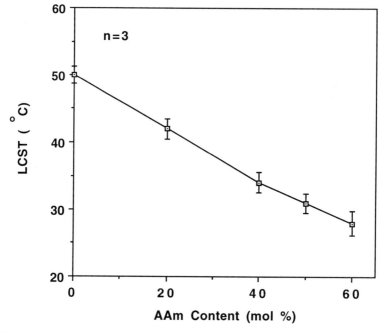

Figure 1. LCST of poly (DMAEMA-co-AAm) solutions as a function of AAm content. (n is the number of experiments)(Reproduced with permission from ref 8. Copyright 1997 John Wiley & Sons, Inc.)

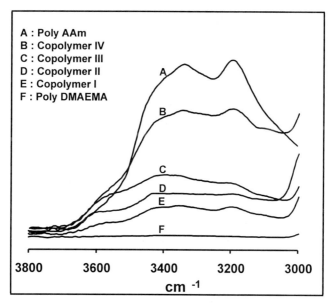

Figure 2. The N-H stretching region of the infrared spectra of copolymers. (Reproduced with permission from ref 8. Copyright 1997 John Wiley & Sons, Inc.)

may be characterized by a rather broad band centered at approximately 3340 cm⁻¹, which is assigned to N-H groups that are hydrogen bonded to C=O groups. The non-hydrogen-bonded N-H stretching is barely detectable as a weak shoulder at about 3440 cm⁻¹. With copolymerization of AAm with DMAEMA, a rather broad band centered at approximately 3340 cm⁻¹ is observed, which is assigned to amide groups that are hydrogen bonded to DMAEMA. And there is an increase in the absorbance with the increase of AAm content in the copolymer. As for the amide in AAm, its role as hydrogen bond donor and acceptor has been demonstrated(11-12). And the dimethylamino group in DMAEMA is known to be a powerful hydrogen bond acceptor (13). Therefore, it is reasonable to suggest efficient hydrogen bondings between amide and N,N-dimethylamino groups.

Using a curve fitting procedure which has been previously described (14), we resolved the N-H stretching region into three components. Two of the curves correspond to the non-hydrogen-bonded and hydrogen-bonded N-H stretching modes and the third corresponds to the Amide B mode which is attributed to phonon vibration involving the intense carbonyl vibrations in Fermi resonance with the N-H stretching fundamental vibrations (15). Table II presents curve-fitting results of the non-hydrogen-bonded and hydrogen-bonded bands. With the increase of AAm content in the copolymer, the area ascribed to the hydrogen-bonded N-H stretching band increases significantly. Conversely, the area of the non-hydrogen-bonded N-H stretching band remains relatively constant. These results strongly suggest efficient hydrogen bondings between amide and N,N-dimethylamino groups.

In their classical study of the LCST of water-soluble polymers, Taylor and Cerankowski proposed as a general rule that the LCST should decrease with increasing hydrophobicity of the polymer (16). As the AAm content in the copolymer increases, the formation of hydrogen bonds between amide and N,N-dimethylamino groups increases. This hydrogen bond protects N,N-dimethylaminoethyl groups from exposure to water resulting in a significant hydrophobic contribution to the LCST. Similar results were reported by Chen and Hoffman (17). This leads to the shift of LCST to lower temperature.

Temperature-sensitive Behavior of Polymer Gel. Figure 3 shows the temperature dependence for the equilibrium swelling of a copolymer gel composed of poly DMAEMA-co-AAm. Transition temperatures between the shrunken and swollen state were shifted to lower temperatures with the increase of AAm content in the gel network. Additionally, swelling change between 20 and 40 °C was decreased. This is attributed to hydrogen bonding in the copolymer and its hydrophobic contribution to the LCST.

A novel semi-IPNs system has been developed to provide a composite gel which shows more rapid and significant temperature-sensitivity. This system consists of two chemically independent polymers in which the proportions and properties of both polymers can be independently varied. The first network consists of SA which provides the crosslinked network. The others are poly DMAEMA-co-AAm which impart temperature-sensitivity into the polymer network.

SA is a polysaccharide which is obtained from marine brown algae. It forms gel in the presence of divalent ions at concentrations of > 0.1 % w/w (18). Because of its remarkable properties of gelation, SA has been extensively exploited and studied in detail in the context of various specific applications or areas. Based on this property of SA, the solution mixture of SA and poly DMAEMA-co-AAm was treated with calcium chloride to form a gel matrix. It could be expected that SA formed a gel matrix (calcium alginate) and poly DMAEMA-co-AAm chains entangled through the calcium alginate gel matrix, resulting in semi-IPNs (19-21). We decided to use Copolymer II in the preparation of composite gels because its crosslinked gel showed drastic swelling change around 34 °C (see Figure 3).

Table II. Curve-Fitting Results of the N-H Stretching

Code	non-hydrogen-bonded N-H			hydrogen-bonded N-H			Amide B		
	V_{cm}^{-1}	$W_{1/2 cm}^{-1}$	Area	V_{cm}^{-1}	$W_{1/2 cm}^{-1}$	Area	V_{cm}^{-1}	$W_{1/2 cm}^{-1}$	Area
Poly DMAEMA	3408.6	144.8	0.7	3311.2	246.9	2.7	3177.3	218.2	2.5
Copolymer I	3428.8	182.9	20.8	3332.4	123.5	6.7	3251.2	129.4	8.6
Copolymer II	3437.1	182.4	22.0	3334.1	226.5	25.8	3194.7	185.2	19.8
Copolymer III	3438.3	90.6	28.1	3334.6	197.2	120.8	3187.0	136.0	74.7
Copolymer IV	3439.6	90.4	22.8	3342.7	206.3	163.9	3182.9	144.4	42.7
Poly AAm	3448.0	90.2	15.0	3349.7	291.3	187.5	3192.7	122.3	36.0

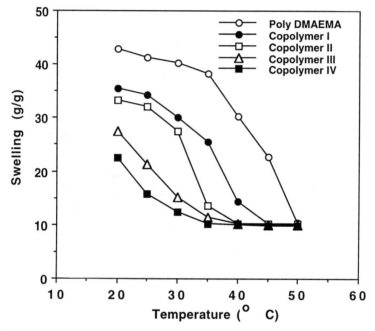

Figure 3. Temperature dependence for the equilibrium swelling of copolymer gels.

Figure 4 shows the temperature-sensitive swelling behavior of composite gels with the variation of Copolymer II content in the gel network. With the increase of Copolymer II content in the gel network, swelling change in response to temperature change was increased. Minimal increase in swelling change in response to temperature change was observed above 50 wt% of Copolymer II. Composite gels contained 1-70 wt % of Copolymer II. If the Copolymer II content was more than 70 wt %, the collapse of the polymer network was observed because of a significant decrease of crosslinking density. Therefore, a composite gel with 50 wt% of Copolymer II content was selected as a model polymer network.

In alkaline aqueous solution, calcium alginate network disintegrates into SA (decoupling) (22). Therefore, the crosslinking density of a polymer network might be regulated by the control of decoupling time (23). Figure 5 shows the swelling change as a function of decoupling time. The swelling increase was observed with the increase of decoupling time. This is due to the decrease in the crosslinking density caused by the decoupling. To observe the change of crosslinking density qualitatively during the decoupling, the calcium ion concentration in the polymer network was assayed by atomic absorption spectroscopy. Figure 6 showed that the concentration of calcium ion in the polymer network decreased with the increase of decoupling time. This indicates that the crosslinking density can be regulated reproducibly by the control of decoupling time.

An optimized composite gel, (5/5, w/w) sodium alginate/Copolymer II with 3 minutes of decoupling, was stored in distilled-deionized water for 1 month and the pH change of the aqueous medium was measured to observe Copolymer II leaking from the composite gel. The leakage of copolymer II can change the pH of the aqueous media because the pH of the 0.5 wt% aqueous solution of Copolymer II is 9.7. However, the lack of pH change of the aqueous medium was observed indicating the minimal leakage of Copolymer II from the gel matrix.

Figure 7 shows the swelling behavior of polymer gels in response to pulsatile temperature changes. Firstly, the temperature was maintained at 38 °C for equilibrium and decreased to 20 °C. The rapid swelling was observed in the composite gel comparing with that of the copolymer gel. In the case of copolymer gel, the conformation change of Copolymer II strain in response to temperature change may be restricted by crosslinking. However, this restriction on the conformation change of Copolymer II may be reduced in the case of the composite gel because of its semi-IPNs structure. Reversible swelling changes were observed for both polymer gels. A more rapid and significant swelling change was observed in the composite gel.

Release rate change from the gel in response to pulsatile temperature changes between 20 and 38 °C is shown in Figure 8. Higher release rates were observed at 38 °C and reduced release rates were observed at 20 °C. The drug release pattern was reversible and this was in accordance with the temperature dependence for the bulk swelling change of the composite gel.

Conclusions

Poly DMAEMA-co-AAm exhibited temperature-induced phase transitions resulting from both polymer-water and polymer-polymer interactions. A hydrophobic contribution to the LCST caused by the intra/intermolecular interaction via hydrogen bonds is a major factor to control the phase transition. Based on the temperature-sensitive behavior of poly DMAEMA-co-AAm, two types of polymer gels were prepared. The composite gel showed a rapid and significant swelling change in response to a pulsatile change. This is due to the semi-IPNs structure of the composite gel resulting in the increased degree of freedom of poly DMAEMA-co-AAm in the composite gel comparing with that in the copolymer gel. The pulsatile release pattern of hydrocortisone in response to a pulsatile temperature change was observed. This was

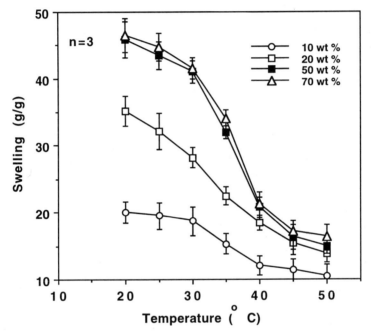

Figure 4. Temperature-sensitive swelling behavior of composite gels as a function of Copolymer II content. (n is the number of experiments)

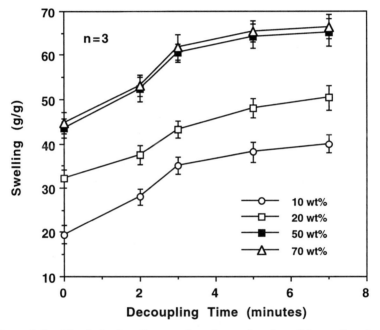

Figure 5. Swelling behavior of composite gels as a function of decoupling time with the variation of Copolymer II content at 25 °C. (n is the number of experiments)

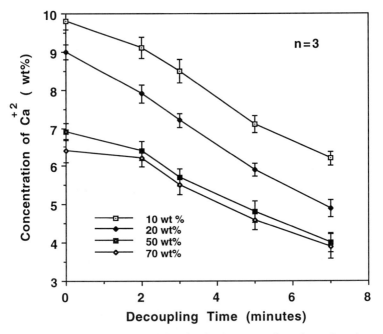

Figure 6. The concentration of calcium ion in the composite gel as a function of decoupling time with the variation of Copolymer II content. (n is the number of experiments)

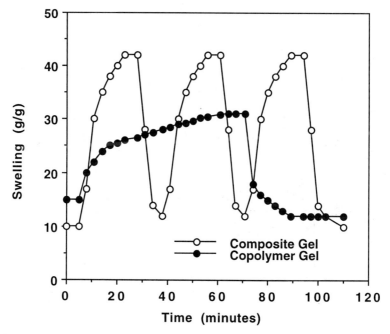

Figure 7. The swelling behavior of composite gel and copolymer gel in response to pulsatile temperature change.

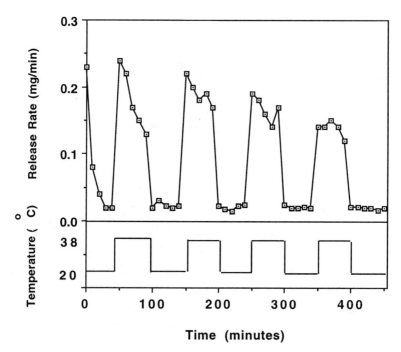

Figure 8. Release change of hydrocortisone from the composite gel in response to pulsatile temperature change.

29

in accordance with the temperature dependence for the bulk swelling change of polymer gels.

Literature Cited

1. Hoffman, A. S.; Afrassiabi, A.;Dong, L. C. *J. Controlled Release* **1986** 4, 213.
2. Bae, Y. H.;Okano, T.;Kim, S.W. *Pharm. Res.* **1991**, 8, 531.
3. Bae, Y. H.;Okano, T.;Kim, S.W. *Pharm. Res.* **1991**, 8, 624.
4. Yoshida, R.;Sakai, K.;Okano, T.;Sakurai, Y. *Polym. J.* **1991**, 23, 1111.
5. Yoshida, R.;Sakai, K.;Okano, T.;Sakurai, Y.;Bae, Y. H.;Kim, S. W. *J. Biomater. Sci. Polym. Ed.* **1992**, 3, 243.
6. Ilman, F. ; Tanaka, T. ; Kokufuta, E. *Nature*, **1991**, 349, 400 .
7. Aoki, T.; Kawashima, M.; Katono, H.; Sanui, K.;Ogata, N.; Okano, T.; Sakurai, Y. *Macromolecules*, **1994**, 27, 947.
8. Cho, S. H.;Jhon, M. S.;Yuk, S. H.;Lee, H. B. *J. Polym. Sci. B; Polym Phys* **1997**, 35, 595.
9. Coleman, M. M.;Shrovanek, D. J.;Hu, J.;Painter P. C. *Macromolecules* **1988**, 21, 59.
10. Shrovanek, D. J.;Howe, S.E.;Painter P. C.;Coleman, M. M. *Macromolecules* **1985**, 18, 1676.
11. Painter, P. C.; Graf, J.; Colemann, M. M. *J. Chem Physics*, **1979**, 92, 6166.
12. Abe, K.; Koide, M. *Macromolecules*, **1977**, 10, 1259.
13. McCormick, C. L.; Blackmon, K. P.;Elliott, D. L. *Polymer* **1986**, 27, 1976.
14. Coleman, M. M.;Lee, K. H.;Shrovanek, D. J.;Painter P. C. *Macromolecules* **1986**, 19, 2149.
15. Moore, W. H.;Krimm, S. *Biopolymers* **1976**, 15, 2439.
16. Taylor, L. D.; Cerankowski, L. D. *J. Polym. Sci., Polym. Chem. Ed.*, **1975**, 13, 2551.
17. Chen, G.;Hoffman, A. S. *Nature* **1995**, 373, 49 .
18. Yalpani, M. In Polysaccharide; Yalpani, M. Eds.;Elsevier:New York, 1988, pp 107.
19. Yuk, S.H.;Cho, S. H.;Lee, H. B. *Pharm. Res.* **1992**, 9, 955.
20. Yuk, S.H.;Cho, S. H.;Lee, H. B. *J. Controlled Release* **1996,** 37, 69.
21. Yuk, S. H.;Cho, S. H.;Shin, B. C.;Lee, H. B. *Eur. Polym. J* . **1996,** 32 101.
22. Kwok, K. K.;Groves, M. J.;Burgess, D. J. *Pharm. Res.* **1991**, 8, 341.
23. Kim, S. R.;Yuk, S.H.; Jhon. M. S. *Eur. Polym. J.* **1997**, 33, 1009.

Chapter 3

Pulsatile Drug Delivery Based on a Complexation–Decomplexation Mechanism

A. M. Lowman[1] and Nicholas A. Peppas

School of Chemical Engineering, Purdue University, West Lafayette, IN 47907-1283

Copolymer networks of poly(methacrylic acid) grafted with poly(ethylene glycol) exhibit pH-dependent swelling behavior due to the reversible formation/dissociation of interpolymer complexes. Because of the complexation/decomplexation phenomena, these gels exhibit large changes in their structure and are able to deliver drugs at varying rates depending on the pH of the environmental fluid. The effects of copolymer composition and the environmental pH on the network structure and the drug release characteristics were studied. The average network mesh was 3 times larger in gels swollen in neutral or basic media than in acidic solutions in which complexation occurred. The largest changes in network structure were observed in gels containing nearly equimolar amounts of methacrylic acid and ethylene glycol. Drug diffusion coefficients, determined through solute release experiments, varied by two-orders of magnitude between the uncomplexed and complexed states.

Hydrophilic, crosslinked polymers which exhibit swelling behavior dependent on their environment have been identified for use in drug delivery applications. The ability of these materials to exhibit rapid changes in their swelling behavior in response to changes in environmental conditions, such as pH, temperature or ionic strength, lend these materials favorable characteristics as carriers for proteins or peptides (1-9). In these systems, the drug is distributed throughout the polymer system, and under conditions of swelling, the drug is able to diffuse out of the polymer network. However, in environments not conducive to swelling, the drug will be entrapped within the network. Using this type of carrier, drugs could be delivered in a pulsatile manner or one in which the drug is released at desired rates for specified intervals to maintain the drug concentration in the body at therapeutic levels for longer time periods (1-10).

[1]**Current Address:** Department of Chemical Engineering, Drexel University, Philadelphia, PA 19104.

Copolymer networks of poly(methacrylic acid-g-ethylene glycol) (P(MAA-g-EG)) have the ability to respond to changes in their environmental conditions. These gels exhibit pH dependent swelling behavior due to the presence of acidic pendant groups and the formation of interpolymer complexes between the ether groups on the graft chains and protonated pendant groups(5,11-16). In these covalently crosslinked, complexing P(MAA-g-EG) hydrogels, complexation results in the formation of temporary physical crosslinks due to hydrogen bonding between the PEG grafts and the PMAA pendant groups (16). The physical crosslinks are reversible in nature and dependent on the pH and ionic strength of the environment. Thus, the number of crosslinks, both chemical and physical, the effective molecular weight of the polymer chains between these crosslinks, M_e, and the end-to-end distance of the polymer chains between these crosslinks or mesh size, ξ, are strongly dependent on the pH and ionic strength of the surrounding environment. In acidic media, such systems are relatively unswollen due to the formation of the intermacromolecular complexes. In basic solutions, the pendant groups ionize and the complexes dissociate.

Complexing hydrogels exhibit drastic changes in their mesh size over small changes of pH. The ratio of the effective hydrodynamic diameter of the drug, d_h, to the network mesh size is critical in determining the ability of the drug to diffuse through the network (17-18). In complexed gels, this ratio decreases significantly from the uncomplexed case. Because of their nature, P(MAA-g-EG) hydrogels would be ideal as carriers of drugs, peptides and proteins (5,11-16). In this work, we have investigated the use of P(MAA-g-EG) hydrogels as pulsatile drug delivery systems.

Experimental Section

Hydrogel Synthesis. The copolymer networks were prepared by free-radical solution polymerization of vacuum distilled methacrylic acid (MAA, Aldrich Chemical Co., Milwaukee, WI) and methoxy terminated poly(ethylene glycol) monomethacrylate (PEGMA, Polysciences, Warrington, PA). The addition of the PEGMA macromonomer resulted in the formation of a network structure containing a PMAA backbone and PEG tethers or grafts. The monomers were mixed in ratios ranging from 1:1 to 4:1 MAA/EG repeating units. The monomer solutions were diluted to 50% by weight of the total monomers by the addition of a 1:1 by weight mixture of ethanol and deionized water. Tetraethylene glycol dimethacrylate (TEGDMA, Polysciences, Warrington, PA) was added as the crosslinking agent to yield a crosslinking density of X = 0.75.

Nitrogen was bubbled through the reaction mixture for 30 minutes to remove dissolved oxygen, a free-radical scavenger, from the solutions. The monomer mixture was stirred under nitrogen and the redox initiator pair, sodium metabisulfite and ammonium persulfate (Mallinckrodt Chemical Inc., Paris, KY), was added in the amount of 2 % total monomer in a nitrogen atmosphere. The mixture was poured between flat plates to form films of 0.9 mm thickness. The monomer films were sealed under nitrogen and allowed to react for 24 hours at 37° C. The ensuing hydrogels were rinsed in distilled water for one week to remove any toxic, unreacted monomer or crosslinking agent, uncrosslinked oligomer chains and initiator.

Characterization of the Network Structure. Following the rinsing, polymer disks were dried under vacuum at 37° C. The dry volumes were calculated using a buoyancy technique. Dry disks of each type were placed in 100 ml of dimethyl glutaric acid buffers ranging in pH from 3.2 to 7.6. The ionic strength of each solution was adjusted to 0.1 M by the addition of NaCl. To determine the equilibrium swelling characteristics, the polymer remained in buffer solution until the volume remained constant (± 5%) for 48 hours. The swelling solution was changed daily for each sample. The polymer volume fraction in the swollen state, $v_{2,s}$, was determined as the volume of dry polymer divided by the swollen polymer volume.

Additionally, rubber elasticity experiments were performed using an automated materials testing system (Instron model 4301, Park Ridge, IL). Gel samples, swollen to equilibrium in DMGA buffers (I = 0.1 M) ranging in pH from 3.2 to 7.4, were cut into strips of width 0.6 mm. The grips of the Instron and the sample were submerged in 37°C buffer solution during the experiment. The polymer samples were elongated at 2 mm/min to reach a maximum elongation of 10%. The tensile stress was calculated for the gels as the force per unstretched, swollen cross-sectional area.

Drug Loading. Drug loading was accomplished by swelling polymer discs (dry thickness of 0.7 mm) in pH 6.8, DMGA buffered, saline solutions containing 1 g/l proxyphylline (Sigma Chemical Co., St. Louis, MO) or vitamin B_{12} (Sigma Chemical Co., St. Louis, MO) at 37° C. To ensure an equilibrium distribution of drug in the gel, the disks remained in the solutions until the concentration of drug in solution and the gel weight remained constant for 3 days. Drug concentrations were monitored using a UV spectrophotometer (Perkin-Elmer Lambda 10, Norwalk, CT). The maximum absorbance of the drug solutions was detected at 273 nm for proxyphylline and 359 nm for vitamin B_{12}. Typical loading times were 5 days for proxyphylline and 7 days for vitamin B_{12}. Following loading, the disks were rinsed in deionized water to remove any drug attached to the external surface and dried under vacuum at 37°C.

Drug Release Studies. Drug release experiments were performed using 1 L dissolution cells. Initially dry, drug loaded disks were placed DMGA buffered saline solutions of specified pH at 37°C. The solutions were stirred at constant rates to ensure that the gels did not adhere to the glass and that both sides of the gel remained exposed to the fluid. Samples were taken at fixed intervals and the drug concentrations were determined using a UV spectrophotometer.

Results and Discussion

The ability of a drug to diffuse through a crosslinked polymer network is dependent on the degree to which the gel swells. In pH-responsive, P(MAA-g-EG) hydrogels, the formation of interpolymer complexes resulted in the presence of temporary physical crosslinks due to hydrogen bonding between the PEG grafts and the PMAA pendant groups. Thus, the degree of crosslinking and the effective molecular weight between crosslinks of these systems decreased due to interpolymer complexation in these gels. Additionally, the correlation length or mesh size was reduced dramatically as more temporary, physical crosslinks were introduced into the system as a result of complex formation. These physical crosslinks were reversible in nature and dependent on the pH of the environment. As a result, the ratio of the hydrodynamic size of the solute to the network mesh size varied dramatically over small pH changes resulting in changes in the diffusion coefficient of the drug in the swollen networks, $D_{3,12}$ (Figure 1).

Equilibrium Swelling. As a hydrogel swells, the polymer chains between crosslink points are elongated, and the network mesh size or correlation length is increased allowing for greater solute permeation in the material. The equilibrium swelling of P(MAA-g-EG) hydrogels is shown in Figure 2. In equilibrium swollen complexation networks, the degree of network swelling is dependent on the number of chemical and physical crosslinks present in the system. Accordingly, the equilibrium swelling ratio was strongly dependent on the pH of the surrounding environment. For the case of equimolar amounts of MAA and EG at low pH values, the degree of complexation was high and the polymer volume fraction in the gel in the swollen state, $v_{2,s}$, was almost 0.70. However, as the pH of the swelling solution increased above pH = 4.6, the complexes began to dissociate and the backbone chains extended resulting in a significant decrease in the equilibrium polymer volume fraction in the gel. The highly swollen, uncomplexed gels contained less than 5% polymer as more water was incorporated into the more loosely crosslinked structure.

The composition of the copolymer significantly affected the degree of swelling of P(MAA-g-EG) hydrogels. In regions of low pH, lower than pH = 4.6, all of the PEG-containing gels exhibited some degree of interpolymer complexation stabilized by hydrogen bonding between protonated pendant acid groups and the ether groups of the PEG graft chains. The largest amounts of complexes and physical crosslinks formed in gels containing equimolar amounts of MAA/EG at low pH. As the amount of MAA in the gels was increased, fewer interpolymer complexes formed in the networks due to the presence of the excess MAA and the gels were able to swell more due to decreased amounts of semi-permanent crosslinks. For solutions of pH greater than 4.6, the acidic pendant groups in all of the gels began to ionize. As the pendant groups were ionized, the number of complexes present in the network structure decreased. As the gel ionized, the complexes dissociated, resulting in a significant decrease in the degree of crosslinking. Additionally, due to the presence of the ionized pendant groups, a large swelling force was generated because of electrostatic repulsions between ionized groups and the development of a large osmotic pressure. As a result, the gel was able to swell to a great extent.

Network Mesh Size Analysis. Because of the complexation/decomplexation phenomena in the P(MAA-g-EG) gels, the mesh size of the networks varied significantly over the pH range of interest. The moduli of the hydrogels for small deformations (less than 10%) were obtained in solutions of differing pH. Based on rubber elasticity theories, the effective molecular weight between crosslinks, $\overline{M}_e$, was calculated from the tensile modulus (16,19-21). Using these data, the mesh sizes were calculated as a function of pH by determining the end-to-end distance of the polymer chains between the crosslinks, both covalent and physical (22).

$$\xi = \left(\frac{2C_n \overline{M}_e}{M_o} \right) \ell \, v_{2,s}^{-1/3} \tag{1}$$

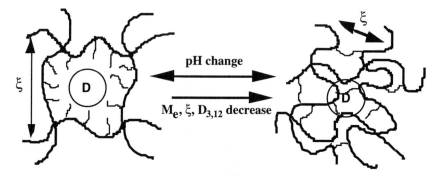

Figure 1. The effect of complexation on the correlation length, ξ, the effective molecular weight between crosslinks, M_e, and the drug (D) diffusion coefficient, $D_{3,12}$, in crosslinked, graft copolymer networks.

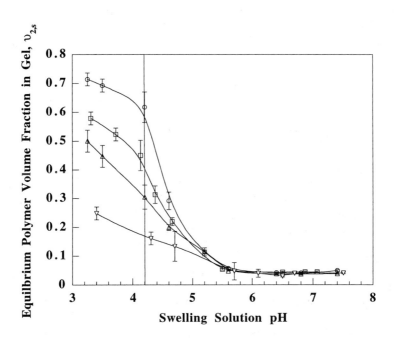

Figure 2. Equilibrium polymer volume fraction in P(MAA-g-EG) hydrogels swollen in buffer solutions (at constant ionic strength, I = 0.1 M) at 37° C plotted as a function of the swelling solution pH for gels with PEG graft chains of molecular weight 1000 containing varying ratios of EG/MAA (1 (o), 3/7 (□) and 1/4 (▲)) and (The value for pure PMAA is denoted by (▽)).

In this equation, C_n (14.6 for PMAA) (*23*) is the polymer characteristic ratio, M_o is the molecular weight of the individual repeating unit and ℓ is the carbon-carbon bond length.

The average network mesh size or correlation length was dramatically affected by the pH of the swelling solution (Figure 3). In low pH solutions in which complexation occurred, the network mesh sizes for P(MAA-g-EG) hydrogels were as low as 70 Å due to the presence of the quasi-permanent, physical crosslinks. However, as the pH was increased the complexes dissociated and the polymer chains elongated resulting in an increase in the network mesh size by a factor of the 3 to almost 210 Å. More importantly, assuming ideal networks, the available area for diffusion is approximately equal to the square of the mesh size. Thus, there existed 9 times greater area for diffusion in the uncomplexed gels (pH greater than 4.6) than the complexed gels (pH less than 4.6).

Drug Release Studies. The important parameter in evaluating the potential of a particular gel to serve as a carrier for a particular drug is the ratio of the effective molecular size (hydrodynamic diameter, d_h) to the network mesh size. Because the network mesh size varies widely over small pH changes, P(MAA-g-EG) hydrogels can be used to alter the diffusional characteristics of various sized solutes based on the pH of the swelling medium.

In order to study the size-exclusion characteristics of these networks, the release of two solutes of differing molecular size, proxyphylline (molecular weight 238 and $d_h = 4.3$ Å) and vitamin B_{12} (molecular weight 1355 and $d_h = 17$ Å), from complexed and uncomplexed gels was studied at 37° C. The diffusion of a solute through a swollen polymer is significantly hindered by the presence of crosslinks as well as physical obstructions such as entanglements, crystallites or polymer complexes (*16-18*). In the complexed gels the rate of drug diffusion was significantly lower than in uncomplexed networks due to the presence of the physical crosslinks., as indicated by the data in Figure 4.

The fractional release of proxyphylline in buffers of varying pH from P(MAA-g-EG) hydrogels is shown Figure 4a. The slowest release was from the gels swollen in the lowest pH solutions. For gels in the most complexed state (pH = 3.5), only 25% of the drug was released from the gel in four hours. Under these conditions, the mesh size of the gel was substantially reduced and the diffusion of the drug was most hindered. As the pH of the environment fluid was increased, the rate of drug release also increased. For gels in which no complexation occurred, 75% of the drug was released in the four-hour time period. Under these conditions, the mesh size of the gel was increased by a factor of three and the diffusion of drug from the gel occurred at a significantly faster rate.

The release of vitamin B_{12} from PEG-containing gels swollen in different pH buffer solutions is shown in Figure 4b. The release behavior of the gel was strongly dependent on the pH of the surrounding fluid due to complexation in the gels. Under conditions of greatest complexation (pH less than 4.8), only 10% of the drug was released from the gel in a four hour period. However, as the pH of the solution was raised to 4.8, a significant increase in the rate of drug release was observed. Under these conditions, the amount of drug released was almost four times greater than the amount released from the gel swollen at pH = 4.6. This was due to the sharp change in the complexation behavior over this pH range. For these networks, the pK_a of the

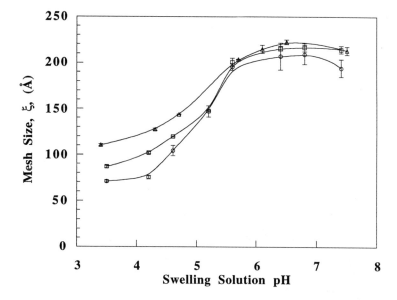

Figure 3. Network mesh size in P(MAA-g-EG) hydrogels swollen in buffer solutions (at constant ionic strength, I = 0.1 M) at 37° C plotted as a function of the swelling solution pH for gels with PEG graft chains of molecular weight 1000 containing varying ratios of EG/MAA (1 (**o**) and 1/4 (**□**)) (The data for PMAA hydrogels are represented by (**▲**)).

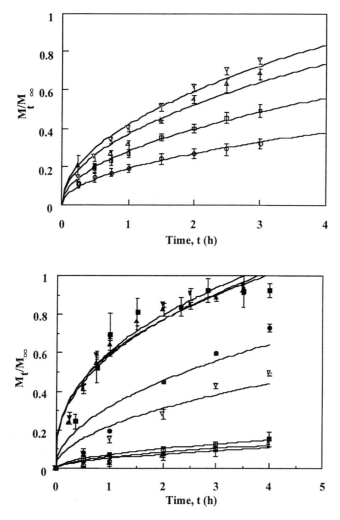

Figure 4. Controlled release (from gels containing PEG grafts of molecular weight 1000 and a 1/1 ratio of EG/MAA) of (a) proxyphylline in DMGA buffered saline solutions (I = 0.1) of pH = (o) 3.5, (□) 4.6, (▲) 5.6 and (▽) 6.8.and (b) vitamin B_{12} in DMGA buffered saline solutions of pH = (o) 3.5, (□) 4.2, (△) 4.6, (▽) 4.8, (●) 5.2, (■) 6.0, (▲) 6.6 and (▼) 7.4 at 37° C.

gel was 4.68. Therefore, the gels swollen at pH = 4.6 contained a significant amount of physical crosslinks. However, at the slightly higher pH, significant ionization occurred causing complex dissociation. In these gels, the mesh size was signicantly larger allowing for significantly faster release of the drug.

As the pH was raised to 5.2, the complexes continued to dissociate allowing for the rate of release to increase. Above pH = 5.2, all of the complexes dissociated and the rate of release from the networks was nearly constant above this point and approximately 80% of the drug was released from these gels.

In order to compare the release kinetics of the P(MAA-g-EG) gels, drug diffusion coefficients, $D_{3,12}$, were calculated using the solution to Fick's second law for diffusion in planar geometries assuming constant diffusion coefficients. At short times, the solution can be approximated (24) as:

$$\frac{M_t}{M_\infty} = \frac{4}{\delta}\left(\frac{D_{3,12}t}{\pi}\right)^{1/2}$$
(2)

In this expression, M_t is the amount released from the gel at any time, M_∞ is the total amount released at infinite time, and δ is the half thickness of the gel. The release data were analyzed using equation 2 and the complexation dependent diffusion coefficients were calculated for the release of the drugs from the hydrogels (Table I).

In the complexed gel, more crosslinks were present to hinder the transport of the solute through the network and the diffusion coefficients for both drugs were decreased significantly. The transport of the larger molecular weight solute, vitamin B_{12}, was more significantly affected by complexation than proxyphylline due to the increased ratio of solute diameter to the network mesh size. In comparison to the case of transport in the highly swollen, uncomplexed networks, the diffusion coefficient for vitamin B_{12} in the complexed material was greater than an order of magnitude less while the proxyphylline diffusion coefficient was only reduced by a factor of 2.

Table I. Ratio of the diffusion coefficients for solutes of varying molecular size in complexed and uncomplexed P(MAA-g-EG) hydrogels containing a 1:1 molar ratio of MAA/EG and graft PEG chains of molecular weight 1000 swollen at 37° C.

Drug	Hydrodynamic Radius, r_h (Å)	r_h/ξ_{com}	r_h/ξ_{uncom}	D_{com}/D_{uncom}
proxyphylline	2.2	0.032	0.011	0.57
vitamin B_{12}	8.5	0.12	0.044	0.096

Pulsatile Release of Vitamin B_{12}. Because the diffusion coefficient and release rate of vitamin B_{12} varied so significantly in the complexcd and uncomplexed gels, the pulsatile release of the drug was studied (Figure 5). For the first 24 hours of the release of vitamin B_{12} from P(MAA-g-EG) gels swollen in pH = 3.2 buffer solutions

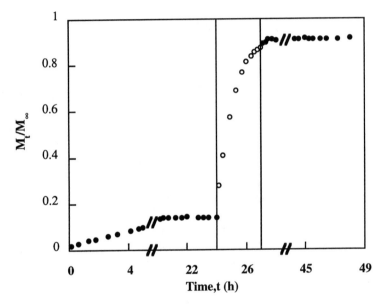

Figure 5. Pulsatile release of vitamin B₁₂ from gels containing PEG grafts of MW 1000 and a EG/MAA molar ratio of 1/1 in DMGA buffered saline solutions (I = 0.1) of pH of (o) 7.1 and (●) 3.2 at 37° C.

was examined. Over the 24 hour period, only 15% of the drug was released. Under these conditions, the gels were in the highly complexed state resulting in the very slow release of the drug from the network.

However, when the hydrogels were placed in environments with pH above the network pK_a (pH = 7.4 buffer solution), the complexes dissociated, allowing the network to swell and rapid drug release to occur. In this case, 60% of the remaining drug was released from the more highly swollen, uncomplexed gel in only 2 hours. When the gels were returned to an acidic fluid after the 2 hour swelling period, the complexes immediately reformed resulting in network collapse. In this case the release was halted due to rapid gel collapse as a result of the formation of interpolymer complexes.

Conclusions

The formation of complexes due to macromolecular associations in polymer networks has a significant effect on the structure and properties of the material. The effects of complexation on the network structure in P(MAA-g-EG) gels and the diffusional characteristics of solutes in the networks were studied. In acidic media, complexes formed in these networks resulting in a more highly crosslinked structure. The diffusion coefficients of solutes and the amount of fluid that the gels could imbibe in the complexed state was significantly reduced due to the presence of additional physical crosslinks. The degree to which complexation occurred varied with the copolymer composition and the molecular weight of the PEG chains grafted to the polymer backbone. The presence of complexes in these gels also provided mechanical stability and reduced the available area for diffusion in the networks. In neutral or basic media, the complexes dissociated allowing the gel to swell to significantly higher degrees. In these gels, the diffusion of solutes in the network was found to occur at significantly higher rates than in the complexed membranes.

Additionally, the complexation/decomplexation mechanism was reversible in nature and exploited to change the release rate of vitamin B_{12} based on the environmental pH. Because these materials were so effective in altering the diffusion rates of moderately sized solutes, these materials could be even more effective for the pH-dependent, pulsatile release of large molecules, particularly proteins and peptides.

Acknowledgments

This work was supported by a grant No. 43331 from the National Institutes of Health.

Literature Cited

1. Hoffman, A. S.; Afrassiabi, A.; Dong, L. C. *J. Controlled Release* **1986**, *4*, 213.
2. Kim, S. W. In *Advanced Biomaterials in Biomedical Engineering and Drug Delivery Systems;* Ogata, N.; Kim, S. W.; Feijen, J.; Okano, T., Eds.; Springer: Tokyo, 1996; 125-133.
3. Peppas, N. A. *J. Bioact. Compat. Polym.* **1991**, *6*, 241.
4. Peppas, N. A. In *Pulsatile Drug Delivery;* Gurny, R.; Juninger, H. E.; Peppas, N. A., Eds.; Wissenschaftliche Verlagsgesellschaft: Stuttgart, 1993; 41-56.
5. Peppas, N. A.; Klier, J. *J. Controlled Release* **1991**, *16*, 203.

6. Siegel, R. A. In *Pulsed and Self-Regulated Drug Delivery*; Kost, J., Ed.; CRC Press: Boca Raton, Florida, 1990; 129-155.
7. Siegel, R. A.; Falamarzian, M; Firestone, B. A.; Moxley, B. C. *J. Controlled Release* **1988**, *8*, 179.
8. Vakkalanka, S. K.; Brazel, C. S.; Peppas, N. A. *J. Biomater. Sci., Polym. Ed.* **1996**, *8* 119.
9. Yoshida, R.; Uchida, K.; Kaneko, Y.; Sakai, F.; Kikuchi, A.; Sakurai, Y.; Okano, T. *Nature* **1995**, *374*, 240.
10. Peppas, N. A. and Langer, R., *Science* **1994**, *263*, 1741.
11. Bell, C. L.; Peppas, N. A. *Adv. Polym. Sci.* **1995**, *122*, 125.
12. Bell, C. L.; Peppas, N. A. *Biomaterials* **1996**, *17*, 1203.
13. Bell, C. L.; Peppas, N. A. *J. Controlled Release* **1996**, *39*, 201.
14. Bell, C. L.; Peppas, N. A. *J. Biomater. Sci., Polym. Ed.* **1996**, *7*, 671.
15. Klier, J.; Scranton, A. B.; Peppas, N. A. *Macromolecules* **1990**, *23*, 4944.
16. Lowman, A. M.; Peppas, N. A. *Macromolecules* **1997**, *30*, 4959-4965.
17. Harland, R. S.; Peppas, N. A. *Colloid Polym. Sci.* **1989**, *267*, 218.
18. Peppas, N. A.; Rheinhart, C. T. *J. Membr. Sci.,* **1983**, *15*, 275.
19. Flory, P. J. *Principles of Polymer Chemistry;* Cornell University Press: Ithaca, NY, 1953.
20. Treloar, L. R. G. *The Physics of Rubber Elasticity;* Oxford University Press: Oxford, 2nd ed., 1958.
21. Peppas, N. A.; Merrill, E. W. *J. Appl. Polym. Sci.* **1977**, *21*, 1763.
22. Peppas, N. A.; Barr-Howell, B. D. In Hydrogels in Medicine and Pharmacy, Vol. 1; Peppas, N. A., Ed.; CRC Press: Boca Raton, FL, 1986; 28-55.
23. Brandup, J.; Immergut, E. H. *Polymer Handbook;* Wiley: New York, 9th ed., 1989.
24. Crank, J. *The Mathematics of Diffusion;* Oxford University Press: Oxford, 1956.

Chapter 4

Gel-Coated Catheters as Drug Delivery Systems

Stevin H. Gehrke[1], John P. Fisher, James F. McBride,
Stephen M. O'Connor, and Huajie Zhu

Department of Chemical Engineering, ML 0171, University of Cincinnati,
Cincinnati, OH 45221–0171

Commercially available poly(ethylene oxide) [PEO] and poly (acrylic acid) [PAA] gel-coated balloon angioplasty catheters are investigated for their use as local drug delivery systems in terms of gel/solute interactions, solute loading, and release kinetics. The solutes investigated are methylene blue, methyl orange, acetaminophen, and ovalbumin which model cationic, anionic, nonionic, and protein compounds, respectively. Solute interactions with gel coatings are found to be consistent with observations with bulk gels. Oppositely charged gel coatings and solutes exhibit ion exchange. Gel coatings and solutes with the same charge demonstrate Donnan ion exclusion; exclusion is reduced upon the addition of salt to the solution. Nonionic interactions, such as size exclusion, are also seen. Loading of proteins in PEO-gel coatings can be approximately doubled with the addition of soluble dextran to the loading solution. Release of solutes from gel coatings is diffusion limited, though resistance may be due to the boundary layer as well as the gel.

A catheter is both a diagnostic and a clinical tool used in a variety of medical procedures. Catheters are often coated with a hydrogel to facilitate their movement through the circulatory system; the gel coating is intended to act as a lubricant. It has been proposed that a hydrogel-coated catheter can also function as a drug delivery system (1). The drug would be absorbed by the hydrogel and then released to a localized site, using the catheter as a vehicle. The drug delivery system could be used while the catheter performs a separate function. For example, a hydrogel-coated angioplasty balloon can both mechanically open an artery and treat the condition that caused the occlusion with drugs (1). The goal of this paper is to correlate the behavior of gel coatings with bulk gels; do they act in the same or different manners? This will be investigated in two areas : (1) the interactions and loading of gel coatings with ionic and nonionic solutes and (2) the release of solutes by gel coatings.

[1]Current Address: Department of Chemical Engineering, Kansas State University, Manhattan, KS 66506–5102.

Principles for Gels Used in Drug Delivery Systems

Hydrogels are excellent materials for drug delivery. The high water content of hydrogels generally enhances their biocompatibility while enabling the inclusion of proteins and other solutes. The key parameter for describing hydrogels is the swelling degree (Q): the ratio of a gel's wet to dry weights. The diffusion coefficient, partition coefficient, and release kinetics of a gel all depend upon Q and thus swelling is a central parameter for drug delivery *(2)*. This work focuses upon solute interactions with gels, enhanced loading of solutes in gel coatings using a technique recently developed for bulk gels, and the release kinetics of solutes from gels.

Solute Interactions. The partition coefficient (K) of a gel-solution system is a measure of a solute's relative affinity for the gel versus the solution. K is defined as the equilibrium ratio of solute concentration in a gel to that in the solution. K larger than one indicates a higher solute concentration in the gel; K between one and zero indicates a higher solute concentration in the solution. By its definition, K cannot be less than zero. A large K is often desired for drug delivery because it allows a gel to absorb a significant amount of a loading solution's solute.

The relative charge on the gel and solute can influence interactions and therefore K's magnitude. A gel and solute of opposite charge generally exhibit considerable levels of ion exchange and thus K is significantly greater than one. Donnan ion exclusion is seen with a gel and solute of like charge; thus, K is less than unity. When either the gel or the solution are nonionic, K can be dominated by size exclusion ($0 <$ K <1) or hydrophobic interactions (K > 1). These ionic and nonionic phenomena are discussed below.

The charge on both the gel and the solute is typically a function of solution pH; Table I outlines this effect. For example, the weak acid poly(acrylic acid) [PAA] gel will exhibit nonionic character at pH 2 and anionic behavior at pH 12. PAA's sorption of a strong base, such as methylene blue, is therefore a function the pH of the loading solution. Theory predicts that significantly more methylene blue will be absorbed by PAA at high pH than at low pH. Clearly a good understanding of solute interactions and the ionic character of the gels and solutes can be used to manipulate K as desired.

Table I: Gel and solute ionic character under high and low pH conditions

	Example	Net Charge pH2	pH12	Gel Coating Tested	Solute Tested
Weak Acid	-COOH	0	-	PAA	none
Strong Acid	-SO$_3$H	-	-	none	Methyl Orange
Weak Base	-NH$_2$	+	0	none	none
Strong Base	-NR$_3$ +	+	+	none	Methylene Blue
Nonionic	-ROR-	0	0	PEO	Acetaminophen

Ion Exchange. Ion exchange allows for the exchange of an ionic solute for a similarly charged counterion that is associated with a polyelectrolyte gel. The system of an anionic gel (containing ionizable carboxylic acid groups) and the chloride salts of a cationic solute (X^+) provides an example of the ion exchange process, as illustrated in equations 1 and 2.

$$\text{at pH 2} \quad RCOOH \; + \; X^+Cl^- \quad \leftrightarrow \quad RCOO^-X^+ \; + \; H^+Cl^- \quad (1)$$

$$\text{at pH 12} \quad RCOO^-Na^+ + \; X^+Cl^- \quad \leftrightarrow \quad RCOO^-X^+ \; + \; Na^+Cl^- \quad (2)$$

When the gel has been equilibrated with a low pH environment, the anionic gel is in its hydrogen form; exchange can then take place with the cationic solute. The reaction is reversed and solute is released with the addition of excess acid (Equation 1). When exposed to a high pH environment, the gel is placed in its anionic form. Once again, exchange can take place with the cationic solute; subsequent release from such a system can be triggered with the addition of excess salt (Equation 2).

Donnan Ion Exclusion. The Donnan effect results in the exclusion of ions from a polyelectrolyte gel when the ion and gel possess the same charge. The exclusion is a consequence of the electroneutrality that must exist in both the gel and the solution. Donnan ion exclusion will be observed in a system where some ions are restricted to a fixed location while others can distribute freely throughout the system. The fixed ions of a gel's polymer chains versus the mobile ions of the surrounding solution are one such system.

Consider the example of an anionic gel in water *(3)*. The gel has a fixed negative charge and associated cationic counterions. An anionic solute tends to move down its concentration gradient into the gel. However, to remain electroneutral, the solute's cations must move along with the anions. But the cations prefer to remain in the external solution due to the high concentration of the gel's cationic counterions. Hence, the solution's anions are effectively excluded from the gel for the sake of the electroneutrality of both the gel and the solution. When a salt is placed into the solution, Donnan ion exclusion will be reduced. The salt will increase the solution's anion and cation concentration relative to that which exists in the gel, thus allowing both the cations and the anions to move into the gel together. Quantitative relationships for this effect have been developed *(4)*.

Nonionic Interactions. A nonionic constituent in a gel/solute system will not be influenced by the phenomena of ion exchange and Donnan ion exclusion. This will be true in an ionic gel-nonionic solute system, a nonionic gel-ionic solute system, or a nonionic gel-nonionic solute system. In these cases, nonionic interactions, especially size exclusion and hydrophobic interactions, dominate *(5)*.

Size exclusion is the exclusion of solutes from a gel due to the relatively large size of the solute when compared to the available openings in the gel's polymer network. In a gel, a solute's conformation is restricted by the polymer chains; in free solution, no such restrictions exist. Hence, solutes prefer the solution to the gel. Thus, solute configuration is a consideration as well as molecular volume. For example, a long, thin cylindrical solute will be more excluded than a spherical solute of the same molecular weight. Various theories have been developed to quantify this effect *(6)*. When size exclusion effects dominate a system, K can be expected to be less than one.

Ideal size exclusion is not observed in many cases due to the presence of hydrophobic interactions: the complex interactions between the hydrophobic regions of a gel's polymer chains and the hydrophobic regions of solutes. These interactions are most pronounced when the solute is hydrophobic and the gel is of low swelling degree (i.e., it is also relatively hydrophobic). Hydrophobic interactions can result in high loading of solute in the gel (K > 1) and often cause deviations from the results expected from size exclusion theory *(7)*.

Enhanced Loading. Large solutes such as proteins tend to be excluded from gels due to size exclusion. Thus delivery of therapeutic quantities of proteins by gels can be difficult due to the minimal levels of loading which can be achieved by sorption.

Techniques have been developed, however, to enhance the loading in bulk gels based on an analogy with aqueous two phase extraction *(8-13;* Gehrke, S. H., Uhden, L. H., McBride, J. F., *J. Controlled Release*, in press). In general, two polymers dissolved in a common solvent will separate into two phases, each enriched in a single polymer. A solute dissolved in such a biphasic system will often distribute unevenly between the polymer phases. By manipulating solution composition (concentration, pH, salt type) the partitioning between the two phases can be made much different than one. This concept has been successfully extended to a gel-polymer solution system and thus proved that enhanced loading of bulk gels is feasible *(8-13)*.

Release Kinetics. The release kinetics of solute from a gel-coated catheter can vary greatly but the kinetics generally must fall within a certain range to achieve the desired therapeutic result. If release is too fast, most solute will be lost prior to the localized delivery; if release is too slow, the catheter must remain at the site of delivery for an unreasonable length of time. Characterization of the release kinetics can allow for considered recommendations about the manufacture and use of the delivery system. Therefore, work is presented to model the movement solutes through gels.

Release can be influenced by a number of factors, notably diffusion in the gel and in the boundary layer. Ion exchange kinetics have also been seen to influence release rate *(14)*. If the release of solute is diffusion-limited and the gel coating is modeled as a thin sheet, the effective diffusion coefficient (D) can be determined from equation 3 *(15)*:

$$\frac{M_t}{M_\infty} = 1 - \sum_{n=0}^{\infty} \frac{8}{(2n+1)^2 \pi^2} \exp\left\{\frac{-D(2n+1)^2 \pi^2 t}{4L^2}\right\} \tag{3}$$

$$
\begin{array}{lll}
\text{where} & M_\infty & = \text{total solute released as } t \to +\infty \\
& M_t & = \text{solute released through time t} \\
& M_t / M_\infty & = \text{fraction of solute released at time t} \\
& D & = \text{effective diffusion coefficient} \\
& t & = \text{time} \\
& L & = \text{gel thickness}
\end{array}
$$

D is determined by a least squares fitting of the calculated release curve to the experimental data.

Experimental

Catheters and Solutes. Two commercially available gel coated balloon angioplasty catheters are investigated: Hydroslide and Bioslide. Hydroslide, produced by the Boston Scientific Corporation (Natick, MA), is a poly(acrylic acid) [PAA] coated catheter; it is anionic when ionized at higher pH. Bioslide, manufactured by SciMed (Maple Grove, MN), is coated with a nonionic gel made from poly(ethylene oxide) [PEO].

Four model solutes (cationic, anionic, nonionic, and protein) are examined. Methylene blue [MB] (320 Da) is a cationic dye, while methyl orange [MO] (327 Da) is an anionic dye. Both solutes were used in the ion exchange/exclusion studies. Acetaminophen (151 Da) is a nonionic solute used in the nonionic interaction studies. Ovalbumin (45000 Da) is a model protein used to test the enhanced loading concept developed for proteins.

Technique. Studies of solute interactions, enhanced loading concept, and release kinetics were performed in the same general manner. The gel-coated catheter is first soaked in a pretreatment solution which, for example, may put a polyelectrolyte gel in its acidic form. The catheter is then placed in the loading solution, allowing the gel to take up solute from the surrounding solution. The catheter may be quickly rinsed in water to remove any excess solute from the surface of the gel-coating. Finally, the catheter is placed in a solution, for a fixed amount of time, into which the solute desorbs.

In the solute interaction and enhanced loading work, equilibrium solution concentration is measured. Solution concentration is measured with an UV/visible spectrophotometer for low molecular weight solutes and a HPLC equipped with a UV detector for the protein. Due to the small amounts of solute released, uncertainty in this measurement dominated the propagation of error estimates reported in the tables. Reproducibility of selected experiments was better than the error propagation estimates reported indicate, however.

Release of solute from the gel-coating was examined by dipping the loaded catheters into a sequence of buffer solutions over different periods of time. Total solute release *versus* time data was generated. Effective diffusion coefficients were obtained by fitting equation 3 to the release data using an Excel spreadsheet program.

Results

Solute Interaction Study. The interactions of a gel and solute of opposite charge, anionic PAA and cationic methylene blue (MB^+), are presented in Table II. The carboxylic groups of the PAA gel change their ionic character with different catheter pretreatments. At pH 2 the groups are protonated ($-RCOOH$) while at pH 12 the groups are in their sodium ion form ($-RCOO^-Na^+$). The gel coating is highly loaded with methylene blue after both pretreatments. (The level of loading reported is accomplished after 5 minutes of solute sorption. Loading was seen to continually increase with time, possibly due to solute aggregation.) No release of solute is observed when the catheter is dipped in water. Release of the bound methylene blue is triggered by acid or 0.5 N NaCl (data not shown), consistent with the ion exchange hypothesis.

Table II: Investigation of gel and solute of opposite charges: methylene blue [MB^+] (cationic) loading in a poly(acrylic acid) gel-coated catheter (anionic)

Catheter Pretreatment	Loading Solution	Release Solution	Total Solute Release (nmol)	% Release into Water
pH 12 NaOH	10^{-3} M MB^+ in water	pH 2	400 ± 28	0
pH 2 HCl	10^{-3} M MB^+ in water	pH 2	280 ± 40	0

The interactions of similarly charged gel and solute, anionic PAA and anionic methyl orange (MO^-), are also investigated (Table III). No loading from a methyl orange solution is seen. However, loading of methyl orange is observed when 0.1 N NaCl is added to the methyl orange solution, as expected with the Donnan ion exclusion hypothesis. The solute releases completely in water, also as expected, since it is not bound within the gel. These results contrast with those in Table II for oppositely charged gel and solute in respect to both loading and release. Though loading is achieved with the addition of salt, sorption of solute is still much less than

Table III: Investigation of gel and solute of similar charge: methyl orange [MO⁻] (anionic) loading in a poly(acrylic acid) gel-coated catheter (anionic)

Catheter Pretreatment	Loading Solution	Release Solution	Total Solute Release (nmol)	% Release into Water
pH 12 NaOH	10^{-3} M MO⁻ in water	pH 12	0	0
pH 2 HCl	10^{-3} M MO⁻ in 0.1 M NaCl	pH 12	6.7 ± 0.5	100

that seen when the gel and solute are oppositely charged. Also, water releases the unbound solute, not the change in pH as when the gel and solute are oppositely charged.

PEO-gel coated catheters and various solutes are used in the investigation of nonionic solute effects (Table IV). First, the interaction of methyl orange (anionic, 327 Da) and PEO gel (nonionic) is examined. The nonionic character of the PEO gel eliminates the need for catheter pretreatment. Methyl orange is loaded into the PEO gel using a water and 0.1 M NaCl solution. Either acidic and basic solutions will allow the solute to desorb from the gel, as there is no specific gel/solute interaction. Release level is similar under all conditions. Complete release of solute is also observed when the catheter is dipped in water. The level of release is significantly less than seen in Table II for oppositely charged gel and solute. Second, the interaction of methylene blue (cationic, 320 Da) and PEO gel is examined. Again, solute is loaded using a water and 0.1 M NaCl solution; acidic and basic both solutions allow release. Release level is similar under all conditions, yet higher than the release of methyl orange. Third, acetaminophen (151 Da) is loaded into a PEO gel using two concentrations of loading solution; water is used as the release solution. The solute is completely released under these conditions. Release levels are similar to methyl orange when similar loading concentrations are used. Acetaminophen loading increases by nearly 10 times when its loading solution is ten times more concentrated.

Table IV: Loading of methylene blue (cationic), methyl orange (anionic), and acetaminophen (nonionic) in a nonionic PEO gel-coated catheter

Solute	Solute Molarity	Loading Solution	Release Solution	Total Solute Release (nmol)	% Release into Water
Methyl Orange	1×10^{-3}	water	pH 12	1.2 ± 2	100
	1×10^{-3}	0.1 M NaCl	pH 12	1.6 ± 2	100
	1×10^{-3}	water	pH 2	1.4 ± 2	100
	1×10^{-3}	0.1 M NaCl	pH 2	7.0 ± 2	100
Methylene Blue	1×10^{-3}	water	pH 12	5.7 ± 1	57
	1×10^{-3}	0.1 M NaCl	pH 12	9.9 ± 1	57
	1×10^{-3}	water	pH 2	6.1 ± 1	66
	1×10^{-3}	0.1 M NaCl	pH 2	6.3 ± 1	75
Acetaminophen	1×10^{-2}	water	water	12 ± 6	100
	1×10^{-3}	water	water	1.6 ± 2	100

Enhanced Loading Study. Table V shows the results of the enhanced protein loading study. Loading of ovalbumin in the PEO gel catheter is accomplished with a pH 2.2 citrate buffer (0.01 M HCl). Release of the ovalbumin is accomplished by soaking in a pH 6.8 phosphate buffer. A similar level of ovalbumin sorption is achieved when 0.1 M KI is added to the loading solution. In contrast, ovalbumin loading and release is doubled upon the addition of 9 wt % dextran to the loading solution. The increase occurs regardless of the presence of the KI salt. The same catheter is used in all trials in Table V, but reuse of the catheter was shown to give reproducible results (data not shown). According to heuristics developed for two phase aqueous extraction, the presence of salt as well as the salt type is expected to affect solute partitioning and thus, for the polymer solution / gel case, solute loading. Further work on the effect of salt is required. However, additional studies show that the addition of 9 wt % soluble PEG does not increase loading.

Table V: Enhanced protein loading of poly(ethylene oxide) gel-coated catheter with solution additives. Loading solution consists of pH 2.2 citrate buffer (0.01 M HCl), 23 mg/mL ovalbumin, and indicated additives. Release solution is pH 6.8 phosphate buffer.

Loading Solution Additives	Protein Release (μg)	Loading Enhancement Ratio
None	44	-
0.1 M KI	43	1.0
9 wt % 40,000 Dextran	88	2.0
0.1 M KI + 9 wt % 40,000 Dextran	79	1.8

Release Kinetics Study. The release kinetics methyl orange and ovalbumin from PEO and PAA gel coated catheters were determined (Figures 1, 2, and 3). Effective diffusion coefficients are calculated from this data and the gel coating thickness. However, the gel coatings were too thin to be directly measured, but were estimated to be about 5 μm thick. Thus the calculated D values are only approximate. Complete release of methyl orange from a PAA gel-coated catheter (data not shown) occurs within 3 seconds, too fast to measure release rate. But this implies that D is greater than 1×10^{-6} cm^2/s. The release of methyl orange from a PEO gel-coated catheter is complete within 20 seconds (D = 4×10^{-7} cm^2/s). The effective diffusion coefficient for ovalbumin is 2×10^{-7} cm^2/s for the PAA coating and 1×10^{-7} cm^2/s in the PEO-coated catheter.

Discussion

Solute Interaction Study. The data presented in Table II clearly shows that the ion exchange phenomenon exists for gel-coated catheters. Both the acidic and basic catheter pretreatments successfully load methylene blue into the PAA gel. The acidic pretreatment allows methylene blue (MB$^+$) to exchange with a hydrogen ion (H$^+$), as described in equation 1. The basic pretreatment allows the cationic methylene blue (MB$^+$) to exchange with a sodium ion (Na$^+$), as described in equation 2. Equations 4 and 5 both demonstrate that release can be triggered by shifting the equilibrium to favor the free MB$^+$ through the addition of either excess HCl or NaCl, as observed.

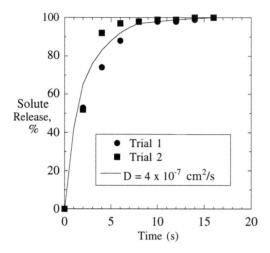

Figure 1 : Diffusional release of methyl orange from a poly(ethylene oxide) gel-coated catheter (Bioslide). The diffusion coefficient is based on an estimated hydrated catheter coating thickness of 5 μm.

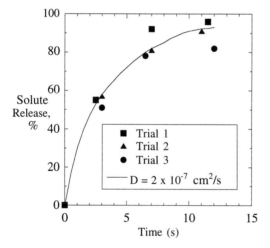

Figure 2 : Diffusional release of ovalbumin from a poly(acrylic acid) gel-coated catheter (Hydroslide). The diffusion coefficient is based on an estimated hydrated catheter coating thickness of 5 μm.

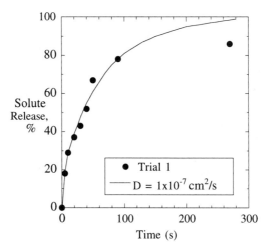

Figure 3 : Diffusional release of ovalbumin from a poly(ethylene oxide) gel-coated catheter (Bioslide). The diffusion coefficient is based on an estimated hydrated catheter coating thickness of 5 μm.

$$\text{at pH 2} \quad \text{RCOO}^-\text{MB}^+ \ + \ \text{H}^+\text{Cl}^- \ \rightarrow \ \text{RCOOH} \ + \ \text{MB}^+\text{Cl}^- \quad (4)$$

$$\text{at pH 12} \quad \text{RCOO}^-\text{MB}^+ \ + \ \text{Na}^+\text{Cl}^- \ \rightarrow \ \text{RCOO}^-\text{Na}^+ + \ \text{MB}^+\text{Cl}^- \quad (5)$$

Data in Table III is consistent with Donnan ion exclusion. When the catheter is soaked in a methyl orange solution at low ionic strength, the anionic PAA gel-coated catheter has no visual evidence of sorption of methyl orange and shows no release of the solute. This clearly indicates that the MO⁻ did not permeate the gel, consistent with predictions of the Donnan theory. This effect is minimized when sodium chloride is added to the loading solution and solute sorption occurs. However, since the dye is not complexed with the gel, such absorbed solute freely diffuses out of the gel when placed in water. Thus the anticipated features of Donnan ion exclusion are observed with the PAA gel-coated catheters.

Table IV shows that a nonionic gel-coated catheter can successfully load any type of solute, regardless of charge. Methyl orange, methylene blue, and acetaminophen are all taken up by the PEO gel-coating. Similar levels of solute release are seen under all loading and release conditions. None of the differences in pretreatment conditions and release solutions key for the ionic gel systems affect loading or release for any of the solutes. Since all of the solutes tested are of small size, similar levels of loading and release are expected and this is observed. However, complete release of methylene blue into pure water does not occur, possibly due to solute aggregation within the gel as observed during loading. The nearly ten times increase in acetaminophen release that accompanies the ten times increase in loading solution concentration is indicative of a partition coefficient which is not concentration dependent (since $K = C_g/C_s$). Since K tends to be independent of concentration for size exclusion while K is highly concentration dependent for ion exchange and ion exclusion, this result supports the hypothesis of ordinary size exclusion for nonionic solutes *(7)*.

Enhanced Loading Study. Table V demonstrates that the loading techniques developed for bulk gels are also applicable to gel-coated catheters. Low ovalbumin sorption by the PEO gel-coating is expected based upon size exclusion theory and this is observed by the modest loading with pure buffer solution. The addition of salt to the buffer solution had no effect upon solute loading. However, the loading of ovalbumin in a PEO-coated catheter is doubled when dextran is added to the solution. Dextran is the key component in the solution additives as expected from the two phase aqueous extraction principle upon which the enhanced loading technique is hypothesized to rely. Without the dextran, a thermodynamic difference in the two phases is not created and thus the ovalbumin partitions between the gel and the salt solution in a conventional fashion (size exclusion). But the presence of the dextran sets up two different environments, which can be adjusted to favor increased loading of ovalbumin into the gel. In contrast, the addition of 9 wt % PEG to the solution does not lead to increased sorption, since the PEG solution provides a similar solution environment to PEG gel. Further study on this concept is required and underway, since some anomalies have arisen; i.e., the presence of the salt and changes in dextran concentration did not alter sorption as expected based on results with bulk gels(16).

Release Kinetics Study. The release of both the methyl orange and ovalbumin solute is shown to be extremely quick due to the thinness of the gel coatings. The solute release from the catheter is complete within 200 seconds for even the slowest case. For *in vivo* applications, the release is too fast; it is impossible to position the catheter properly before nearly all of the solute would be released from the gel.

Release of solutes from both PAA and PEO gel-coated catheters is found to be diffusion-limited, demonstrated by the approximate fit of the theoretical curve to the experimental data (other examples have been obtained in addition to those shown here). Release is characterized by effective diffusion coefficients of approximately 4×10^{-7} cm^2/s for methyl orange and 2×10^{-7} cm^2/s for ovalbumin. These are approximate values as thickness of the gel coating was too thin to be measured accurately. The larger ovalbumin molecule possesses the smaller D as expected. In water, molecules of size similar to methyl orange possess a D of about 5×10^{-6} cm^2/s; ovalbumin is known to have a D = 7.8×10^{-7} cm^2/s *(17)*.

The effective diffusion coefficients estimated are smaller than expected for diffusion through the gel, which implies that solute diffusion may not be limited strictly by movement through the gel, but also through the boundary layer surrounding the gel coated catheter. Although this argument depends in part on thickness measurements, such a layer was in fact visible with colored solutes. This result has important implications for *in vivo* use of the gel as a drug delivery system. In the application, solute must diffuse from the catheter into the arterial wall. Since diffusion in the gel is not necessarily the limiting step, it can be expected that studies of diffusion through the arterial wall will be required for the ultimate success of drug delivery by this method.

Conclusions

The study of gel-coated balloon angioplasty catheters has produced four key conclusions. (1) Ionic and nonionic solute interactions that are observed in bulk gel systems are also seen in gel-coated catheter systems. (2) Enhanced loading techniques that have been previously developed for bulk gels are applicable to gel-coated catheters. (3) The release of solutes by commercially available catheters is too quick for effective drug delivery. (4) The release of solutes by gel-coated catheters can be expected to be influenced by external mass transfer resistance.

It is recommended that catheters be manufactured with thicker gel coatings if catheters are to be used for local drug delivery. The thicker coating will allow for a longer period of time to place the catheter before release is complete as well as allow for delivery of increased doses. Clearly, however, the concept works and the technique is promising. The ability to use standard concepts that are well-established in the literature for bulk gels in the design of hydrogel coatings for catheters will greatly facilitate successful development of commercial devices following the concepts outlined in this paper.

Acknowledgments

This research was supported by the Boston Scientific Corporation, Natick, MA. The authors would also like to thank Dr. Ron Sahatjian and Dr. James Barry for helpful discussions.

Literature Cited

1. Fram, D. B.; Aretz, T.; Azrin, M. A.; Mitchel, J. F.; Samady, H.; Gillam, L. D.; Sahatjian, R.; Waters, D.; McKay, R. G. *J. Am. Coll. Cardiol.* **1994**, *23*, pp 1570-1577.
2. Gehrke, S. H.; Lee, P. I. In *Specialized Drug Delivery Systems*, Tyle, P. Ed.; Marcel Dekker: New York, NY, 1990; pp 333-392.
3. Gehrke, S. H.; McBride, J. F.; O'Connor, S. M.; Zhu, H.; Fisher, J. P. *Polym. Mat. Sci. Eng.* **1997**, *76*, pp 234-235.
4. Yin, Y.L; Prud'homme, R. K. In *Polymers of Biological and Biomedical Significance,* Shalaby, S. W.; Ikada, Y.; Langer R.; Williams, J. Eds.; ACS Symposium Series 540; American Chemical Society: Washington, DC, 1994; pp 157-170.
5. Lund, M. E. M. S. Thesis, University of Cincinnati, Cincinnati, OH, 1996.
6. Schnitzer, J. E. *Biophys. J.* **1988**, *54*, pp 1065-1076.
7. Gehrke, S. H.; Palasis, M.; Lund, M. E.; Fisher, J. P. *Ann. NY Acad. Sci.* **1997**, *831*, pp 179-207.
8. Gehrke, S. H.; Vaid, N. R.; Uhden, L. H. *Proceed. Intern. Symp. Control Rel. Bioact. Mater.* **1993**, *20*, pp 113-114.
9. Gehrke, S. H.; Uhden, L.H.; Schiller, M. E. *Proceed. Intern. Symp. Control Rel. Bioact. Mater.* **1995**, *22* pp 145-146.
10. Gehrke, S. H.; Robeson, J.; Johnson, J. *Biotech. Prog.* **1991**, *7*, pp 355-358.
11. Gehrke, S. H.; Vaid, N. R.; McBride, J. M. *Biotech. Bioeng.* **1998**, *58*, pp 416-427.
12. Gehrke, S. H.; Lupton, E. C.; Schiller, M. E.; Uhden, L.; Vaid, N. U.S. Patent 5,603,955, **1997**.
13. Gehrke, S. H.; Lupton, E. C.; Schiller, M. E.; Uhden, L.; Vaid, N. U.S. Patent 5,674,521, **1997**.
14. Gehrke, S. H.; Cussler, E. L. *Chem. Eng. Sci.* **1989**, *44*, pp 559-566.
15. Crank, J. *The Mathematics of Diffusion, 2nd Ed.*; Oxford University Press: London, UK, 1975; p 49.
16. Uhden, L. H. M.S. Thesis, University of Cincinnati, Cincinnati, OH, 1995.
17. Cussler, E. L. *Diffusion: Mass Transfer in Fluid Systems*; Cambridge University Press: Cambridge, UK, 1984; p 116.

Chapter 5

Novel Ethylene Glycol-Containing, pH-Sensitive Hydrogels for Drug Delivery Applications: "Molecular Gates" for Insulin Delivery

Christie M. Hassan and Nicholas A. Peppas[1]

Biomaterials and Drug Delivery Laboratories, School of Chemical Engineering, Purdue University, West Lafayette, IN 47907–1283

Poly(methacrylic acid-g-ethylene glycol) (P(MAA-g-EG)), poly(N-vinyl pyrrolidone-g-ethylene glycol) (P(NVP-g-EG)) and poly(N-vinyl pyrrolidone-co-methacrylic acid) (P(NVP-co-MAA)) gels were investigated as possible carriers for pH-sensitive release of drugs using the characterisitics of complexation/decomplexation due to hydrogen bonding. Glucose-sensitive P(MAA-g-EG) gels were also investigated for use in novel self-regulated devices that deliver appropriate amounts of insulin in response to changing glucose levels leading to a system/device working with the principle of "molecular gates". These gels were synthesized by first activating glucose oxidase and then polymerizing the gel comonomers in the presence of the activated enzyme. The equilibrium swelling behavior of the gels was examined as a function of pH. At high pH values, the gels swelled to approximately 20 times their dry weights. The swelling/deswelling behavior of the glucose oxidase-containing gels was investigated under varying pH conditions to characterize their dynamic swelling behavior. The glucose oxidase-containing gels exhibited a higher rate of expansion than the non-glucose oxidase containing-gels. Glucose-sensitive gels were capable of releasing 0.5 mg of insulin in 5 minutes and an additional 0.5 mg over 3.5 hours. The release from glucose-sensitive P(MAA-g-EG) hydrogels could be described as Fickian with an initial burst of insulin.

Hydrogels are crosslinked polymers capable of absorbing water because of the presence of hydrophilic functional groups (1) such as -OH, -COOH, -CONH$_2$, -CONH, and -SO$_3$H. From a physicochemical point of view, hydrogels resemble natural tissue and often exhibit good biocompatibility because of their high water content. They have been used in various biomedical applications, including biosensors and drug delivery systems, because of these characteristics (2,3).

Hydrogels can also be sensitive to the conditions of the external environment because of the presence of certain functional groups along the polymer chains. The swelling behavior of these materials may be dependent on pH, temperature, ionic strength, or even glucose concentration (4). In crosslinked networks that contain weakly acidic (anionic) and/or basic (cationic) pendent groups, water sorption can result in ionization of these pendent groups depending on the solution pH and its ionic

---

[1] Corresponding author.

composition. The gel then acts as a semi-permeable membrane to the counterions (5). This influences the osmotic balance between the hydrogel and the external solution through ion exchange, depending on the degree of ionization and ion-ion interactions. For ionic gels containing weakly acidic pendent groups, the equilibrium swelling ratio increases as the pH of the external solution increases (6), while the swelling ratio increases as the pH decreases for gels containing weakly basic pendant groups (7).

There has been increased research interest in the development of glucose-sensitive polymeric systems (8,9) which deliver appropriate amounts of insulin in response to changing glucose levels so as to mimic the natural response of the body. This could lead to better control of blood glucose levels in diabetic patients. This approach involves an enzyme-substrate reaction that results in a pH change and use of a pH-sensitive polymer that responds to the change. Glucose reacts with glucose oxidase forming gluconic acid thus, decreasing the pH of the environment. With the change in pH, the gel swells or collapses depending on the characteristics of the particular polymer of the system. Insulin is released from this system with the change in the size of the pores of the polymer (8).

Complexing Hydrogels for Oscillatory Insulin Delivery

The materials of interest in this research were complexing poly(methacrylic acid-g-ethylene glycol) hydrogels henceforth designated as P(MAA-g-EG). Interpolymer complexes are formed in these hydrogels due to hydrogen bonding (10) between the hydrogens of the carboxylic group of the poly(methacrylic acid) (PMAA) and the oxygens on the ether groups of the poly(ethylene glycol) (PEG) chains. At low pH values, there is sufficient protonation of the carboxylic acid groups causing complexes to form. This results in a collapse of the gel due to increased hydrophobicity in the polymer network. At high pH values, complexes break as the carboxylic groups become ionized. This results in an expansion of the gel as electrostatic repulsion is produced within the network. In our specific hydrogel systems, the two species involved in the complexation are bound together in the same polymer by grafting EG on the PMAA backbone chains. This allows for the reversible formation of complexes at appropriate conditions. Therefore, these materials allow for pH sensitive solute permeation, a property which may be important for drug release (11).

Previous work based on the interpolymer complexation between PMAA and PEG has been reviewed by Osada (12). For example, Osada and Takeuchi (13) examined the effects of treating poly(methacrylic acid) with poly(ethylene glycol). They observed dilation and contraction of the system due to reversible complexation of PMAA with PEG. The effects of the PEG chain length used to treat PMAA membranes were investigated by Osada (14). It was found that PEG with low molecular weight of 600 and 1000 produced rapid but small contractions and easily attained equilibrium. PEG of molecular weight 2000 yielded a rapid and pronounced contraction. PEGs of molecular weight greater than 2000 showed considerable contractions over a long period of time.

Klier et al. (15) were the first to investigate the preparation and characterization of P(MAA-g-EG) networks where grafting of the two structures prevents total dissolution upon decomplexation. Self-associating networks were prepared by the copolymerization of methacrylic acid with poly(ethylene glycol) monomethacrylate in the presence of the crosslinking agent, tetraethylene glycol dimethacrylate. The swelling of the networks depend on swelling solution pH, swelling temperature, copolymer composition, and network structure. Nuclear Overhauser enhancement measurements indicated that graft copolymers of MAA and PEG formed complexes for a wider range of concentration and PEG molecular weights than the 2 ungrafted homopolymers. Copolymer networks were also found to swell to a lower extent than homopolymer networks due to complex formation between PEG and PMAA segments.

Bell and Peppas (*11,16,17*) have done extensive work to characterize this swelling behavior by performing equilibrium swelling studies as a function of pH and oscillatory swelling studies as a function of time and pH. Equilibrium swelling studies indicated that the mesh sizes between the complexed (low pH) and uncomplexed (high pH) states for all samples increased by 96 to 99%. At low pH, the mesh sizes of the various samples were small, less than 10 Å, whereas at high pH in the uncomplexed or expanded state they were in the range of 240 Å to 350 Å. Mesh sizes of the networks were also calculated under oscillatory pH conditions. The mesh sizes responded rapidly to pH changes. Maximum mesh sizes in the expanded states ranged from 11 to 27 Å, whereas in the collapsed states they ranged from 4 to 9 Å.

In the present work, we have reacted glucose oxidase on P(MAA-g-EG) to form a glucose- and pH-sensitive "squeezing" gel. At high concentrations of glucose, the glucose oxidase catalyzed reaction of glucose can produce gluconic acid resulting in a decrease in the pH of the environment around the gel. Glucose-sensitive hydrogels may thus collapse in response to the decrease in pH, and insulin can be squeezed out of the network.

In the present work we also introduce for the first time other grafted hydrogels that may exhibit hydrogen bonding and associated complexation/decomplexation behavior. These gels include poly(N-vinyl pyrrolidone-g-ethylene glycol), henceforth designated as P(NVP-g-EG), and poly(N-vinyl pyrrolidone-co-methacrylic acid), henceforth designated as P(NVP-co-MAA).

Experimental

Synthesis of Glucose-Sensitive Hydrogels. Glucose-sensitive gels of P(MAA-g-EG) were synthesized by activation of the enzyme and polymerization using a procedure adapted from those developed by Valuev and Platé (*18*). These techniques have been discussed before by Hassan et al. (*19*). Briefly, in the enzyme activation step, a buffering solution of sodium carbonate was prepared by dissolving 300 mg of sodium carbonate in 5 mL of deionized water. Next, 0.1 g of glucose oxidase type VII from *Aspergillus niger* (Sigma Chemical Co., St. Louis, MO) was added to the buffer solution. Catalase from *Aspergillus niger* (Sigma Chemical Co., St. Louis, MO) was then added to the solution in the amount of 350 mL. An ice bath was prepared to chill this solution to 4°C. Then 2 mL of acryloyl chloride (Aldrich Chemical Co., Milwaukee, WI) was added. The solution was mixed for 1 hour with a magnetic stirring bar. It was important that the solution remained at a temperature of 4°C or below during this hour because acryloyl chloride reacts vigorously. This glucose oxidase solution served as a solvent in the polymerization reaction.

The monomers used for the preparation of the glucose-sensitive hydrogels were methacrylic acid (Aldrich Chemical Co., Milwaukee, WI) and methoxy-terminated poly(ethylene glycol) monomethacrylate (Polysciences, Inc., Warrington, PA; PEG MW of 400). The methacrylic acid was purified by vacuum distillation at 66°C and 30 mm Hg. A 50:50 ratio of repeating units of MAA to repeating units of PEG was used. Tetraethylene glycol dimethacrylate (TEGDMA) (Aldrich Chemical Co., Milwaukee, WI) was added as the crosslinking agent in the amount of 2 weight % total monomer. In a typical experiment, 8.8 g of methacrylic acid, 1.2 g of PEG monomethacrylate, and 0.1 g TEGDMA were mixed and diluted with 5 g of ethanol and 5 g of glucose oxidase solution. The solution was then placed in a glove box. The glove box was purged with nitrogen for 45 minutes. Nitrogen was then bubbled through the polymerization mixture for 20 minutes to remove oxygen. Next 0.04 g of ammonium persulfate and 0.04 g of sodium metabisulfite were added as the redox initiators. The polymerizing solution was stirred well for 20 minutes and poured into a glass petri dish cover containing 0.7 mm thick teflon spacers. A petri dish bottom was then placed over the solution and spacers so as to create a thin film of polymer between the petri dishes. An additional petri dish cover was then placed on top. The arrangement was sealed with duct tape, removed from the nitrogen glove box, and the mixture was allowed to polymerize at room temperature for 24 hours. Upon completion of the reaction, the ensuing polymer was removed from the petri dish and cut into thin disks of 12 mm diameter. The samples were then stored at $-20^{\circ}C$ to retain enzymatic activity.

The glucose oxidase-containing P(MAA-g-EG) hydrogels were tested for enzymatic activity in a solution of 100 mg of dextrose, anhydrous (J.T. Baker, Inc., Phillipsburg, NJ) per 100 mL of deionized water as discussed before (*19*). The initial pH of the solution was measured with a pH meter (model 399A; Orion Research, Inc., Cambridge, MA). A polymer disk was placed in the solution for 15 minutes and the final pH of the solution was measured. A significant drop in the pH, caused by glucose oxidase catalyzing the reaction of glucose in solution to produce gluconic acid, indicated whether or not the glucose oxidase remained active during the polymerization.

Synthesis of Poly(N-vinyl pyrrolidone-g-ethylene glycol). Poly(N-vinyl pyrrolidone-g-ethylene glycol) hydrogels were synthesized by photopolymerization of methoxy-terminated poly(ethylene glycol) monomethacrylate (Polysciences, Inc., Warrington, PA; PEG MW of 200) and N-vinyl pyrrolidone (Aldrich Chemical Co., Milwaukee, WI) mixtures with NVP:PEG weight ratios of 50:50, 20:80, and 80:20. Tetraethylene glycol dimethacrylate (TEGDMA) was added as the crosslinking agent in the amount of 2 wt% of total monomers. The reaction mixture was then diluted with a 50:50 mixture (by weight) of ethanol and water to a 60:40 monomer:solvent ratio. In a typical experiment, 5 g of NVP, 5 g PEG 200 monomethacrylate, and 0.2 g TEGDMA were mixed and diluted with 3.34 g ethanol and 3.34 g distilled/deionized water. Irgacure 184 (CIBA-GEIGY Corp., Hawthorne, NY) was added as the photo-initiator in the amount of 1 wt% total monomer. Using a pipette, this polymerization solution was then loaded between glass microscope slides with 1 mm thick spacers and exposed to ultraviolet light from an Ultracure 100 lamp (Efos USA, Inc., Williamsville, NY) for 15 minutes.

Synthesis of Poly(N-vinyl pyrrolidone-co-methacrylic acid). Poly(N-vinyl pyrrolidone-co-methacrylic acid) hydrogels were also synthesized by similar photopolymerization techniques. The mixtures contained N-vinyl pyrrolidone and methacrylic acid with NVP:MAA weight ratios of 60:40, 70:30, and 80:20. TEGDMA, ethanol, water, and initiator were all added in the same amounts as in the preparation of the P(NVP-g-PEG) films. A UV exposure time of 12 minutes was used.

Swelling Studies. Equilibrium swelling studies were performed at various pH values to characterize the swelling behavior for glucose-sensitive P(MAA-g-EG) hydrogels. Dimethyl glutaric acid/sodium hydroxide buffer solutions of pH 3.2, 4.0, 4.8, 5.4, 5.8, 6.0, and 7.0 were prepared by combining 100 mL of a 0.1 M solution of 3,3-dimethylglutaric acid (Sigma Chemical Co., St Louis, MO) and appropriate amounts of a 1 N sodium hydroxide solution (Sigma Chemical Co., St. Louis, MO) and diluting to 1 L with distilled/deionized water. The ionic strength was kept constant for all solutions at I=0.1 M by adding appropriate amounts of sodium chloride (Fisher Scientific Co., Fair Lawn, NJ) to each solution during preparation.

Equilibrium swelling studies were also performed with P(NVP-g-EG) and P(NVP-co-MAA) hydrogels over the pH range of 3 to 10.5. The pH of the solution was adjusted by adding sodium hydroxide or hydrochloric acid to distilled/deionized water. Ionic strength was also held constant by adding appropriate amounts of sodium chloride.

Dried samples were weighed and placed in 50 mL of a solution of particular pH at $37^{\circ}C$. The samples were blotted and weighed and the solutions changed daily until the weight of the sample did not change by more than 0.01 g over a 24 hour period.

Varying pH conditions were used for both the glucose-sensitive P(MAA-g-EG) and non-glucose oxidase-containing P(MAA-g-EG) gels to better characterize their dynamic behavior. Dimethylglutaric acid/sodium hydroxide buffer solutions of pH 4.0 and 7.0 were used for these oscillatory experiments. A polymer disk of 0.5 mm thickness was equilibrated in the pH 4.0 solution at 37°C. Then, the disk was placed in the pH 7.0 solution (still at 37°C) for 45 minutes. The sample was blotted and weighed every 5 minutes. After 45 minutes, the sample was placed back into a new pH 4.0 solution for 45 minutes. The sample was again blotted and weighed every 5 minutes. This 90 minute cycle of pH 7.0 to pH 4.0 was repeated 2 additional times.

Insulin Release Experiments. The release of insulin was studied from glucose-sensitive P(MAA-g-EG) hydrogels in response to pH. Disks of 5 mm thickness and 12 mm diameter were equilibrated in a pH 7.0 glutaric acid/NaOH buffer solution at 37°C. An insulin loading solution of 0.2 g/L was prepared by first dissolving 0.02 g insulin (from bovine pancreas, Sigma Chemical Co., St. Louis, MO) in 1 mL of 0.01 M hydrochloric acid solution. The solution was then diluted to 100 mL with pH 7.0 buffer solution. The pH of the solution was adjusted to pH 7.0 by adding sodium hydroxide. An equilibrated disk was then placed in the 0.2 g/L insulin solution for 4 hours with constant stirring. After 4 hours, the disk was rinsed with distilled/deionized water and placed in a pH 4.0 glutaric acid/sodium hydroxide buffer solution in a shaker water bath operated at 37°C. Samples of 3 mL were removed from the solution and then replaced with 3 mL of pH 4.0 buffer solution every 5 minutes for the first 30 minutes and at specific times for the following 3 hours. The ultraviolet absorbances were measured with a UV/Vis Spectrometer (Lambda 10 model, Perkin Elmer, Norwalk, CT) at 274 nm.

Results and Discussion

Synthesis of Hydrogels. Glucose oxidase-containing hydrogels of P(MAA-g-EG) were synthesized by activation of the enzymes, and polymerization. The purpose of the activation of glucose oxidase and catalase was to provide structures that allowed the enzymes to be chemically attached to P(MAA-g-EG) during the polymerization step. In the procedure that was followed, the acrylate group of the acryloyl chloride attached to the enzymes at the site of amines on the enzymes. This formed a peptide bond either at the N-terminus of a polypeptide or on an amino acid side group, allowing the enzymes to maintain activity (18). It was important to maintain the solution at a temperature below 4°C during its 1 hour of mixing because of the vigorous nature of the reaction.

The actual polymerization of the glucose oxidase-containing P(MAA-g-EG) involved adding the activated glucose oxidase/catalase solution to the polymerization solution. The hydrogels produced from this technique appeared to be homogenous and seemed to have polymerized completely. Because of the bright yellow color of the glucose oxidase, the polymer could be examined to determine how the glucose oxidase was attached during the polymerization. Continuous yellow coloring, upon visual inspection, throughout the polymer indicated that the glucose oxidase was evenly distributed. Storage of the enzyme solution before use in the polymerization resulted in non-uniform attachment of the enzymes. A yellow coloring only at the surface of the polymer film indicated that the enzymes were concentrated mostly on the surface and that the glucose oxidase did not attach uniformly in the polymer or may not have attached at all. Because of its large molecular weight, glucose oxidase may have been pushed to the outside of the polymerization solution where it was no longer interfering in the polymerization.

It was also found in this polymerization procedure that it was again necessary to bubble nitrogen through the reaction mixture for at least 20 minutes to get complete polymerization. This polymerization also occurred at room temperature (25°C) whereas the standard P(MAA-g-EG) required a polymerization temperature of 37 °C.

The glucose oxidase-containing hydrogels were tested to determine if the glucose oxidase remained active throughout the activation and polymerization techniques used as described before (*19*).

Using the photopolymerization techniques described, poly(N-vinyl pyrrolidone-g-ethylene glycol) hydrogels were also synthesized. A UV exposure time of 12 minutes was found to produce good films. The adhesive characteristics of the gels were found to increase with higher amounts of NVP. Poly(N-vinyl pyrrolidone-co-methacrylic acid) hydrogels were also synthesized using a UV exposure time of 12 minutes in order to obtain good films. Films produced with higher amounts of NVP became somewhat brittle and took on a white appearance when placed in a solution of water. For this reason, the highest amount of NVP used was 40 wt%.

Swelling Studies. The equilibrium swelling behavior of glucose-sensitive P(MAA-g-EG) hydrogels was investigated over the pH range of 3.2 to 7.0. The weight equilibrium swelling ratio, q, was calculated for each pH value by

$$q = \frac{W_s}{W_d} \tag{1}$$

where W_s is the weight of the swollen sample and W_d is the weight of the dry matrix.

Figure 1 shows the equilibrium swelling behavior of thin disks of the glucose oxidase-containing P(MAA-g-EG) gels. At low pH values of 3.2 and 4.0, the gel was in a collapsed state due to complexation or hydrogen bonding between the carboxylic acid groups of PMAA and the oxygens of the ether groups of the PEG chains. At pH values above 5.8, the gels swelled to approximately 20 times their dry weights due to breakage of most of the complexes. Hydrogen bonds were broken at higher pH values as the carboxylic groups became ionized. As a result, the gel began to swell to a high extent as electrostatic repulsion was increased within the network.

The equilibrium swelling behavior of P(NVP-g-EG) gels in the range of pH 3.0 to pH 9.0 is shown in Figure 2. The weight swelling ratio was found to increase with increasing pH for all NVP:MAA ratios with the exception of the sample containing 50 wt% NVP and 50 wt% PEG 200. The swelling ratio of this sample decreased slightly from pH 3.0 to pH 6.0 and increased at pH 9.0. The maximum change in the swelling ratio with a change of 6 pH units (from pH 3.0 to pH 9.0) was approximately 1.0 g swollen polymer/g dry polymer as shown with the 80 wt% NVP and 20 wt% PEG 200 samples.

Figure 3 shows the equilibrium swelling of P(NVP-co-MAA) over the pH range of 3.0 to 10.5. The gels were in a collapsed state (with a weight swelling ratio of about 1.5) over the pH range of 3.0 to 9.0. As the pH was increased to 10.5, the gels swelled to a high extent. The gels were found to swell to between 14 to 22 times their dry weights. The swelling/syneresis behavior of glucose oxidase-containing P(MAA-g-EG) gels was also studied under varying pH conditions to characterize their dynamic behavior. This behavior was also compared with the swelling/syneresis behavior of non-glucose oxidase-containing P(MAA-g-EG). In these studies, the pH of the environment was changed from pH 7.0 to pH 4.0 every 45 minutes. The weight swelling ratio, q, was calculated for each data point using equation (2). The volume swelling ratio, Q, was then calculated by

$$Q = 1 + \frac{\rho_p}{\rho_w}(q - 1) \tag{2}$$

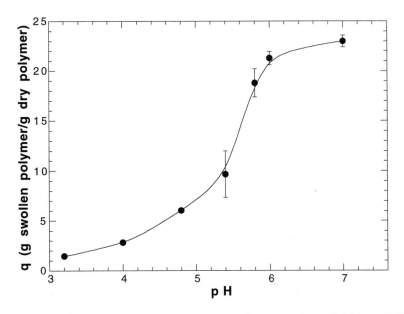

Figure 1. Equilibrium swelling of glucose oxidase-containing P(MAA-g-EG) hydrogels as a function of pH at 37°C.

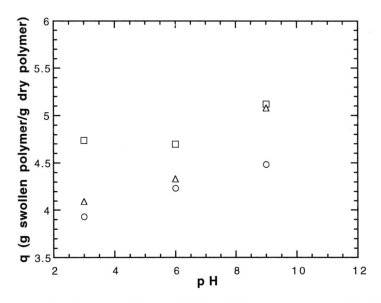

Figure 2. Equilibrium swelling of P(NVP-g-EG) hydrogels with NVP:PEG weight ratios of 20:80 (○), 50:50 (□), and 80:20 (Δ) as a function of pH at 37ºC.

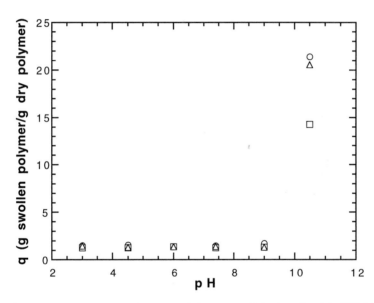

Figure 3. Equilibrium swelling of P(NVP-co-MAA) hydrogels with NVP:MAA weight ratios of 20:80 (○), 30:70 (▢), and 40:60 (△) as a function of pH at 37°C.

where ρ_p is the density of the dry polymer and ρ_w is the density of water. Figure 4 shows the changes observed in the volume swelling ratios of the gels as a function of time under varying pH conditions. In the initial solution of pH 4.0, the gels were in their collapsed state due to complexation. The volume swelling ratios for both the glucose oxidase-containing and non-glucose oxidase-containing gels were around 3.0 cm^3 swollen polymer/cm^3 dry polymer. When the samples were placed into a pH 7.0 solution, hydrogen bonds were broken as the carboxylic acid groups became ionized. This resulted in swelling of the gels.

The swelling ratios increased at a fairly constant rate in the pH 7.0 solution for both gel types. The glucose-sensitive gels exhibit a higher rate of expansion than the non-glucose oxidase containing gels. Because of the presence of the large enzymes, glucose oxidase and catalase, the glucose-sensitive gels had fewer entanglements. Therefore, these gels could expand faster than the non- glucose-sensitive gels. It is also shown in Figure 4 that there is an initial rapid collapse for both types of gels when they were placed in pH 4.0 solutions. The remaining collapse period of 40 minutes at pH 4.0 was rather slow. When the gels were initially placed in the pH 4.0 solutions, complexes formed rapidly causing the gel to collapse quickly. The glucose-sensitive and non-glucose-sensitive gels appeared to have similar collapse (complexation) behavior although their rates of swelling (decomplexation) differed. In both cases, the gels did not return to their initial equilibrium state at pH 4.0 after 45 minutes. This caused the swelling ratios to shift to slightly higher values with each repeated cycle. However, the swelling ratios increased (pH 7.0) and decreased (pH 4.0) by the same relative amount in each cycle. By allowing the gels to remain in a pH 4.0 solution for a longer period of time, the initial equilibrium swelling ratio could be achieved in the repeated cycles.

Insulin Release Studies. The release of insulin was studied from thin disks of glucose-sensitive P(MAA-g-EG) hydrogels. The release behavior was analyzed in terms of drug release models. Ritger and Peppas (20) developed an exponential expression which relates the amount of drug released from a thin polymer film as a function of time,

$$\frac{M_t}{M_\infty} = k\, t^n \tag{3}$$

where M_t is the amount of drug released at time, t, and M_∞ is the total amount of drug released. The kinetic constant, k, depends on the characteristics of the polymer, drug, and medium. The diffusional exponent, n, can be used to identify the time-dependence of the release rate. In Fickian diffusion (n=0.5), polymer relaxation does not affect transport. However, relaxation processes become rate limiting with Case II transport (n=1.0) resulting in drug release that is proportional to time. Both Fickian diffusion and polymer relaxation affect drug release in anomalous transport (0.5 < n < 1.0).

The insulin release behavior from glucose-sensitive P(MAA-g-EG) is shown in Figure 5. There was an initial rapid release of 0.5 mg of insulin from the system in the first 5 minutes. An additional 0.5 mg was then slowly released over the remaining 3.5 hours of the experiment. By performing a mass balance on the loading solution, it was found that approximately 1.2 mg of insulin was loaded into the hydrogel. Therefore, within the first 5 minutes, approximately 40% of the loaded insulin had been released. Slighly over 80% of the insulin had been released after 3.5 hours.

Because of the initial burst effect of insulin observed, Equation (3) could not be used to accurately model the insulin release observed from glucose-sensitive P(MAA-g-EG). The expression was modified to describe this burst effect by,

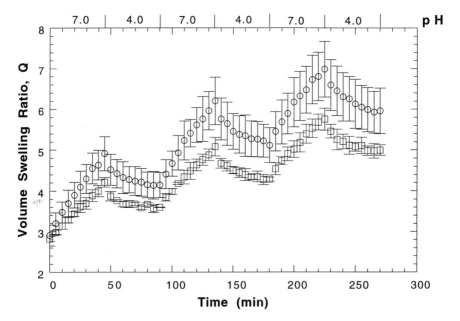

Figure 4. Oscillatory swelling behavior of glucose oxidase-containing (○) and non-glucose oxidase-containing (□) P(MAA-g-EG) hydrogels as a function of time and pH at 37°C, as the gels were swollen at pH values of 7 and 4.

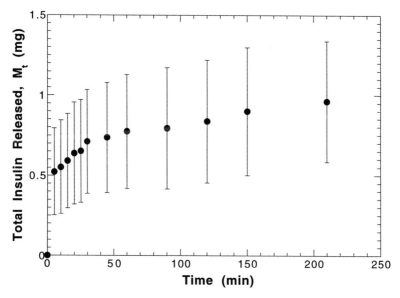

Figure 5. Insulin release from glucose-sensitive P(MAA-g-EG) hydrogels at pH 4.0 and 37°C.

66

$$\frac{M_t}{M_\infty} = \alpha + k\, t^n \qquad (4)$$

where α represents the initial rapid release of insulin, which was assumed to occur instantaneously when the gel was placed in solution. The insulin release behavior was fit to this equation, as shown in Figure 6, where $\alpha=0.39$, $k=0.034$ min$^{-1/2}$, and $n=0.47$. Therefore, the release from glucose-sensitive P(MAA-g-EG) can be described as Fickian with an initial burst of insulin. The initial rapid release of insulin was due to the loading technique used in which the gel was soaked in a concentrated solution of insulin for 4 hours. Therefore, insulin was only able to load at the surface for this small time period. The gels could not be loaded for a longer period of time as the insulin would have started to denature. The release behavior observed can be described as a diffusional process rather than a squeezing process because the loaded insulin was located at the surface of the gel.

Design of an Insulin Release Device. The previous results can be used in a first prototype for the development of an insulin delivery device. Such a system would use the swelling/deswelling of the glucose-oxidase P(MAA-g-EG) gels to prepare a "molecular gates" system.

In this system, any strong porous membrane can be used as a support between an insulin reservoir and the blood. The hydrogen bonding/complexing P(MAA-g-EG) gel can be deposited in the pores as shown in Figure 7. During exposure to high levels of glucose, the glucose oxidase-catalyzed reaction can take place leading to decreased local pH and synersis (shrinkage) of the gels. More specifically, the glucose in the surrounding solution will react in the presence of glucose oxidase to produce gluconic acid according to the reaction:

$$\text{Glucose} + O_2 + H_2O \xrightarrow{\text{Glucose Oxidase}} \text{Gluconic Acid} + H_2O_2$$

Our results have indicated that the glucose oxidase remains active throughout the activation of enzymes and polymerization procedures. Catalase can also be incorporated into the gels. It catalyzes the conversion of hydrogen peroxide to oxygen and water.

$$2\,H_2O_2 \xrightarrow{\text{Catalase}} O_2 + 2\,H_2O$$

The incorporation of catalase is important in order for the glucose oxidase to continue to react in the presence of glucose to produce gluconic acid. In the absence of catalase, the glucose oxidase reaction would be limited because of the disappearance of oxygen in the environment.

This process can lead to an increase in the pore size (opening of molecular gates), thus allowing insulin to be released by diffusion ("molecular gates" principle). As glucose levels decrease by the action of insulin, the local pH increases causing swelling of the gels. This results in a closing of the molecular gates and a membrane that is impermeable to insulin.

Acknowledgments. This work was supported in part by a grant from the Showalter Foundation (Indianapolis). We wish to thank Professor Frank Doyle, III, and Kairali Podual of Purdue University for helpful discussions.

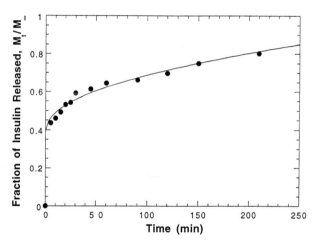

Figure 6. Model fit of fractional insulin release from glucose-sensitive P(MAA-g-EG) hydrogels at pH 4.0 and 37°C.

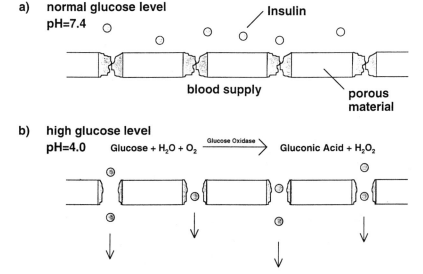

Figure 7. Mechanism of action of the "molecular gate" system: (a) for normal glucose levels, at physiological pH, the gels are swollen and the pores are closed; and (b) for high glucose levels, the glucose oxidase-catalyzed reaction decreases the local pH causing shrinkage of the gels and the diffusion of insulin through the membrane.

Literature Cited
1. Peppas, N.A. *Hydrogels in Medicine and Pharmacy*, CRC Press, Boca Raton, FL 1986.
2. Peppas, N.A. *J. Bioact. Compat. Polym.* **1991**, 6, 241.
3. Peppas, N.A. Langer, R.S., *Science* **1994**, 263, 1715.
4. Peppas, N.A. In *Pulsatile Drug Delivery: Current Applications and Future Trends*; Eds., R. Gurny, H. Junginger and N.A. Peppas; Wissenschaftliche, Stuttgart, 1993, 41.
5. Gehrke, S.H.; Lee, P.I. In *Specialized Drug Delivery Systems*; Editor, P. Tyle; Marcel Dekker, New York, NY, 19 ; 333.
6. Michaeli, I.; Katchalsky, A. *J. Polym. Sci.* **1957** 23, 683-696.
7. Siegal, R.A.; Firestone, B.A. *Macromolecules* **1988** 21, 3254.
8. Heller, J. *Critical Reviews in Therapeutic Drug Carrier Systems* **1993**, 10 253.
9. Albin, G.; Horbett; T.A.; Ratner, B.D. In: *Pulsed and Self-Regulated Drug Delivery*; Editor, J. Kost; CRC Press: Boca Raton, FL, 1990; 159.
10. Bell, C.L.; Peppas, N.A. *Adv. Polym. Sci.* **1995** 122, 125.
11. Bell, C.L.; Peppas, N.A. *Biomaterials* **1996** 17, 1203.
12. Osada, Y. *Adv. Polym. Sci.* **1987** 82, 2.
13. Osada, Y.; Takeuchi, Y. *J. Polym. Sci.: Polym. Lett. Ed.* **1981** 19, 303.
14. Osada, Y. *J. Polym. Sci.: Polym. Lett. Ed.* **1980** 18, 281.
15. Klier, J.; Scranton, A.B.; Peppas, N.A. *Macromolecules* **1990** 23, 4944.
16. Bell, C.L.; Peppas, N.A. *J. Biomater. Sci., Polym. Ed.* **1996** 7, 671.
17. Bell, C.L.; Peppas, N.A. *J. Contr. Rel.* **1996** 39, 201.
18. Valuev, L.I.; Platé, N.A. *Adv. Materials* **1990** 2, 405.
19. Hassan, C.M.; Doyle, F.J.; Peppas, N.A. Macromolecules, in press.
20. Ritger, P.L.; Peppas, N.A. *J. Controlled Release.* **1987** 5, 23.

Chapter 6

Overcoming the Skin Barrier with Artificial Smart Carriers of Drugs

Gregor Cevc

**Medical Biophysics, Klinikum r.d.I., Ismaningerstrasse 22,
The Technical University of Munich, D-81675 Munich, Germany**

Specialized intercellular lipidic seals and the tight packing of cells in the skin prevent molecules greater than a few hundred Dalton from crossing the organ. However, composite, 'intelligent' aggregates, that exert directed pressure on the intercellular junctions in the skin, can open numerous hydrophilic channels through the stratum corneum and carry pharmacological agents, including polypeptides, across the intact permeability barrier after non-occlusive administration. The distinct characteristics of such aggregates, Transfersomes ('carrying bodies'), are the non-uniform and high aggregate deformability (mechano-sensitivity) and the high responsiveness to external transcutaneous gradient(s) (e.g. hydro-sensitivity). Combination of both these properties enables Transfersomes to move through the skin typically between the cell envelope and the skin sealing lipid multilayers. Aggregates without the described characteristics, such as conventional liposomes or mixed lipid micelles, remain confined to the skin surface. The results of in vivo experiments with corticosteroids (hydrocortisone, dexamethasone, triamcinolone), non-steroidal anti-inflammatory agents (diclofenac), and proteins (calcitonin, insulin, serum albumin) in Transfersomes illustrate the main advantages of 'smart carriers': the drug targeting into the skin ($\geq$ 99.9%), the improvement of regional/systemic drug concentration ratio ($10\times$), and the high efficiency of non-invasive macromolecular delivery. In every case, the duration of biological action is also prolonged.

Advancements in materials science over the last decade have prompted the search for corresponding biological and therapeutic applications. The pharmaceutical industry, in particular, seeks new materials (polymer solutions, gels, liposomes or nano-particles suspensions) for more directed, more sustained or better acceptable drug delivery. The holly grail of this research is to find means for non-invasive delivery of large molecules, such as polypeptides and proteins. Oral polymer [1], liposome [2] or micro-emulsion [3, 4] suspensions, or the unusual, e.g. rectal [5], perocular [6], intranasal [7] or dermal [4, 8] applications to date gave unsatisfactory results. This is mainly due to the difficulty of overcoming the underlying biological barrier.

In the skin, the narrowness of passages is a direct consequence of special anatomical and biochemical organization of the organ [9]. The cells in the skin are organized in clustered 'columns'. Cell 'edges' inside each cluster overlap extensively but there is little such lateral overlap between the cells from adjacent clusters. Furthermore, the inter-corneocyte space is sealed with ample lipids, many of which are in the compact, multilamellar and essentially impermeable crystalline phase [10]. The packing of intercellular lipids is less regular and tight in the inter-cluster region than between the corneocytes in an individual group of cells [11].

Two quantitatively different penetration pathways therefore exist in the skin [11]. The first pathway leads between the corneocyte clusters, has a relatively low total area and a low penetration resistance. The second, intra-cluster pathway is more abundant and goes between the laterally overlapping corneocytes within the individual clusters. This latter route frequently traces the irregularities in intercorneocyte lipid lamellae and, even more often, between such lamellae and the neighbouring corneocyte envelopes. Owing to its tortuosity, and due to the high order of lipid structures in the inter-corneocyte region, the intra-cluster pathway has a much higher penetration resistance than the inter-cluster route. The peak in skin resistance is in either case near the stratum corneum conjunctum [11]. Very few entities of appreciable size can trespass this barrier [12, 13].

For controlled drug delivery, to overcome the difficulty of skin crossing, progress must be made in at least two directions: 1) minimize the increase in barrier resistance with the size of the transported entity; 2) counteract the lowering of transport driving force with increasing transportant size, originating from a decreased (molar) drug solubility.

We believe that drug delivery by means of 'smart carriers' with non-Newtonian rheological properties and unusual mechano-sensitivity can solve the problem [14]. However, the design of such carriers fulfilling both abovementioned requirements is a serious challenge. The reason for this is that all drug carriers, in one way or another, rely on aggregation or association and therefore are large. In order to overcome the skin barrier such composite carriers must be made sufficiently adaptable. To be really practial, they should also be useful without gadgets.

In this chapter we discuss how the above listed goals can be reached. We argue that the hetero-aggregates of sufficient size provide the most natural and biologically acceptable solution for overcoming the skin barrier. We present the rationale for designing complex mechano- and hydro-sensitive, self-aggregating mixtures. We demonstrate that the resulting drug carriers have 'smart' properties and can cross the skin spontaneously after non-occlusive administration; we also provide experimental evidence for the notion that this is due to hydrotaxis. Finally, we compare the carrier potential of various kind of aggregates and show how well different drugs are transported into the body by means of ultradeformable lipid vesicles, Transfersomes.

Skin Model

To characterize the essential properties of mechano-sensitive drug carriers we used non-biological skin models. Dehydration stress across the 'artificial skin' barrier was created by keeping one side in contact with a large water reservoir (5 mL); the activity coefficient of water on the other barrier side was fixed by flushing the sample chamber with the air of known relative humidity (30% to ~ 99%): water activity gradient across the barrier was essential for the success of aggregate motion through the narrow pores (cf. figure 2).

The use of artificial skin eliminated the problem of biological tissue variability as well as any interference from the direct interactions between the formulation components and the barrier. Simultaneously, this permitted us to study the pores similar to those believed to be created in the living skin during ultradeformable aggregate transport. To confirm that 'artificial skin' data are relevant for the biological systems we compared qualitatively our present results with those measured previously with the living skin [15, 16, 17].

We first tested the transport of conventional lipid bilayer vesicles across the artificial barrier and made the following observations. As a rule, liposomes were confined to the side of application, independent of the details of administration (cf. figures 1, 2 and 3), when their size appreciably exceeded the diameter of pores in the barrier. This is illustrated in figure 1 for the vesicles which, according to the literature, have elastic energy similar to that of red blood cells, and much higher than the thermal energy, $\kappa > 15RT$ [18, 19]. We therefore conject that conventional vesicles are too stiff to penetrate pores smaller than 2 vesicle diameters. This is true for hydration (cf. o in figs. 2 and 3; dashed line in inset to 2) or hydrostatic (cf. fig. 1, o) pressure as transport driver.

The high hydrophilicity of typical liposome surface, which determines the vesicles hydration energy and makes the bilayer strongly refractory to dehydration [15], does not suffice to pay the energetic price for strong liposome deformation (cf. 5 and [20]): conventional lipid vesicles therefore break rather than cross the micro-porous barrier [21]. Likewise, standard liposomes do not cross the intact skin surface (cf. fig. 2, right) but rather fuse and form extended lipid multilamellae on the skin [22, 23].

When the molar elastic energy of lipid bilayers is lowered below the thermal energy membranes disintegrate spontaneously [20, 14]. Such process of small aggregate formation is often induced by detergents and finally results in complete membrane solubilization, micellization [24]. The mixed lipid micelles are typically much smaller than liposomes and easily cross the pores refractory to the latter (cf. fig. 1, left). However, micelles are incapable of crossing the skin in appreciable quantity in vitro [17] or in vivo (cf. fig. 1, right, and fig. 5). In our opinion, this is due to their inability to widen 'virtual' channels in the skin — which pre-exist in the artificial skin model — to the size permitting the passage.

Transfersomes combine the high dehydration sensitivity of liposomes with the low penetration resistance of micelles [14, 17, 25, 26]. The high and stress-dependent (fig. 1: ●), but size-mismatch insensitive (fig. 3: ●), flow of Transfersomes across the barrier is indicative of certain 'intelligence' of such aggregates.

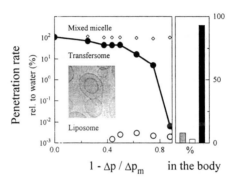

Figure 1: Left: Aggregate transport (in % of the average value measured
with water or mixed micelles (◇)) from a Transfersome (2 w-%, •, $\kappa \sim RT$) or
liposome (2 w-%, ◦, $\kappa \sim 20RT$) suspension as a function of the flow-driving
pressure difference across an artificial skin barrier ($2r_{pore} \simeq 100$ nm). *Inset:* A high magnification cryo-electron micrograph of a large ($r_v \sim 200$ nm),
thermally and hydro-dynamically deformed Transfersome (inset height: 350
nm). Right: the efficacy of tritiurated dipalmitoylphosphatidylcholine transfer across the intact murine skin *in vivo* into the body from a suspension of
mixed micelles containing bile salt (grey column), phosphatidylcholine liposomes (white column) or Transfersomes (black column). (Modified from [26].

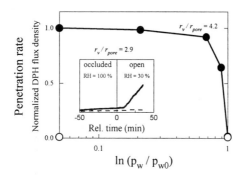

Figure 2: Effect of water activity difference across the barrier $(\ln(p_w/p_{w0}) = \ln(a_w/a_{w0}) \sim \Delta a_w$, 'relative humidity (RH) gradient') on the trans-barrier flow of lipophilic fluorescent marker DPH, diagnostic of aggregate penetration through the narrow pores with $2r_{\mathrm{pore}} = 30$ nm. *Inset:* Time dependence of corresponding aggregate flow.

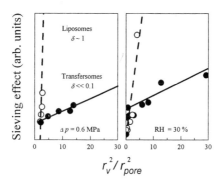

Figure 3: Influence of relative size (r_v/r_{pore}) of liposomes (o) or Transfersomes (•) on the vesicle sieving by narrow pores with an applied pressure (left) or water activity (right) gradient.

The high swelling potential of lipid mixtures which self-aggregate into Transfersomes is suggestive of strong sensitivity to dehydration. These two virtues combined enable Transfersomes to pass the artificial as well as living skin efficiently, only little affected by the narrowness of pores (cf. fig. 3).

Drug Transport in vivo

Owing to their strong sensitivity to the trans-barrier hydration gradient, Transfersomes can increase the specificity of agent delivery into the treated skin and beyond. Transfersomes also spontaneously can carry associated drugs, both small and large, into the body and its blood circulation [16, 17, 26, 27].

Small Agents. Examples of low molecular weight drugs tested to date in combination with Transfersomes include corticosteroids [28], hormones (such as tamoxifen [17]), non-steroidal anti-inflammatory agents (such as diclofenac [16, 17]), antibiotics, etc.. Formulation and administration details can be found in previous publications [15, 25, 26, 28].

The results obtained with corticosteroids are given in figure 4. They illustrate the high degree of regio-specificity ensured by Transfersome-mediated drug transport into the skin: Absolute dexamethasone concentration in the blood or treated skin, as deduced from the -derived radioactivity, changes in the opposite direction with the administered dose, when Transfersomes are used as carriers. If desirable, the agent spill-over from Transfersomes into systemic blood circulation can be nearly completely prevented. The first reason for this is that Transfersomes experience a strong driving force only at the very skin surface and, once they cross the skin, are too big to pass through the blood capillary walls; the other is the reduced total drug amount. For adequate skin treatment, only a small amount of the drug and carrier are required [28].

Diclofenac for the regional treatment is, and should be, used at a higher dose per area than corticosteroids for the skin treatment. This justifies Transfersome application at a higher area-dose with the former drug. Our extensive experiments have shown that diclofenac concentration, mediated by an epicutaneous application of drug-loaded Transfersomes at the dose of 0.3 mg/kg body weight (BW), exceeds the therapeutic drug level, derived from the commercial hydrogel at a much higher applied dose (1.2 mg /kg BW); the bio-distribution of this NSAID from Transfersomes is also shifted towards deeper subcutaneous tissue, which is highly desirable [17]

Peptides and Proteins. Attempts to deliver therapeutic amounts of macromolecules across the skin have been unsuccessful to date [30, 31]. Transfersomes are a breakthrough in this field: different polypeptides (such as calcitonin [29]) or proteins (such as interferon-gamma or serum albumin [33]) administered on the intact skin surface in ultradeformable carriers ultimately give rise to similar drug concentrations in the blood as corresponding subcutaneous injection of the drug in Transfersomes. When used as antigens, large proteins delivered across the skin by means of Transfersomes also give rise to high antibody titers [32, 33].

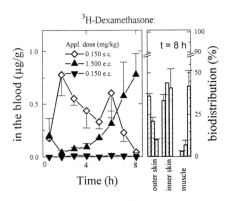

Figure 4: Carrier-mediated regio-specificity of transcutaneous ³H-dexamethasone delivery by means of Transfersomes. (Injection: open symbols or columns; epicutaneously in Transfersomes over similar area: full symbols. 0.015 mg/kg: ▽, left-dashed; 0.15 mg/kg: ◇, hatched; 1.5 mg/kg: △, right-dashed; data from [28].)

The above mentioned results based on radioactivity measurements are substantiated by the outcome of experiments with transcutaneously delivered insulin in Transfersomes (Transfersulin).

Insulin is normally prevented from crossing the skin by its high molecular weight of 5808 Da. When associated with Transfersomes, however, this agent is propelled across the barrier by the carrier.

The representative results of the biological experiments with mice treated with insulin in Transfersomes are presented in the top panel of figure 5. The epicutaneously administered Transfersulin lowers the blood glucose concentration by $20 \cdots 30\%$ within 2 h to 4 h (cf. fig. 5, top: •). In contrast to this, a mixture of insulin with mixed lipid micelles ($\diamond$) or simple liposomes (o) as well as empty Transfersomes (reference and negative control) do not change glucose concentration in the serum significantly over the investigated time period.

Human data reveal similar trends. Simple insulin-loaded liposomes (cf. fig. 5, bottom: o) or the corresponding mixed micelles (data not shown) fail to induce significant hypoglycemia over the first 8 h following the drug administration. Skin pre-treatment with a suspension of mixed micelles and the subsequent use of standard liposomes is also inefficient in our experience. Only when insulin is non-occlusively applied in Transfersomes does significant systemic hypoglycemia evolve. This happens at $t \geq 90 \cdots 180$ min, depending on the detailed carrier composition. The delay is at least partly due to the slower biological action of carrier-associated drug [34] but time required for water evaporation from the skin and drug liberation as well as carrier redistribution under the skin is also important.

Penetration of properly made Transfersomes is sufficiently controllable (see the confidence limits in fig. 5) to ensure the insulin-induced hypoglycemia in mice and humans to be proportional to the applied agent amount in the dose range below 0.2 U Transfersulin s.c./animal and below 0.6 U Transfersulin e.c./animal [26]. In normo-glycemic humans dose linearity is observed below approx. 0.6 U Transfersulin e.c./kg body weight (data not shown). The corresponding linear-regression analyses give correlation coefficients of $R = 0.96$ for mice and of $R = 0.94$ for man. This supports our view that the observed effects are due to the non-invasive delivery of Transfersulin rather than to some other drug- or carrier-independent effect.

Discussion

To date, no unanimous answer exists to the question of how and why an entity as massive as a typical lipid vesicle ($MW \geq 10^6$ Da) could cross the skin [23]. Many disbelieve the basic idea and quote the high permeability barrier of the intact skin, which will only let molecules with a molecular weight lower than 500 Da pass, as the reason for this [13]. Other researchers would have us believe that the epicutaneously applied vesicles disintegrate in the skin and thus permit a small proportion of their components to diffuse into the body [23]. We have previously argued that suitably designed, large but composite, entities trespass the skin if they

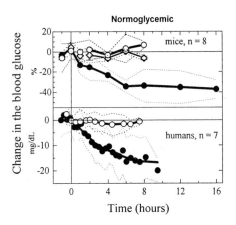

Figure 5: The average glucose concentration change in the blood of mice (upper panel, in %) or humans (lower panel, in mg glucose per dL blood, as a function of time after the epicutaneous administration of insulin (0.45 U/mouse or 0.6 U/kg human body weight) at t=0. (Thick curves: mean values; thin lines: 95 % confidence limits for Transfersomes (●, full), liposomes (○, dashed) and mixed micelles (◇, dotted); for more details see the original ref. [26].)

are deformable enough to squeeze themselves between the corneocytes, through otherwise confining pores [15]. This capability is due to the transient changes in aggregate composition at the sites of greatest stress: the component which sustains deformation better is accumulated while the less adaptable components are diluted at the maximally stressed site. Aggregates of suitable composition, Transfersomes, are therefore self-optimizing: the controlled component flow between the sites of different deformation ensures the high, but spatially variable deformability. A direct consequence of this is the high carrier membrane flexibility and optimum pore-penetration capability. Both translate into the high drug transport efficiency under provision that the Transfersome driving force is strong enough.

Aggregate Transport. Material motion across a barrier is driven by various trans-barrier potential differences. Normally, the chemical potential difference caused by transportant concentration jump across the barrier is thought to be the main such difference. In aggregate transport, however, the trans-barrier gradients not dominated by the transportant concentration prevail. The reason for this is that with increasing aggregation number (n_a) the effective transportant concentration decreases and finally becomes negligeably small. In the description of aggregate transport within the framework of Onsager formalism [17] we therefore only have to consider the 'external', i.e. transportant insensitive potential gradients:

$$J_a = \tilde{L}_{aw}(n_a)(\Delta\mu_a/n_a) + \tilde{L}'_{aw}(n_a)\Delta\mu_{w,i} + \cdots \rightarrow \tilde{L}'_{aw}(n_a)\Delta\mu_{w,i} \qquad (1)$$

such as transportant-independent water activity ($\Delta\mu_{w,i}$) or hydrostatic pressure difference across the barrier.

The phenomenological penetrability factors in eq. 1 are related to the system's resistance to aggregate motion across the barrier. Consequently, the diffusion-describing factor ($\tilde{L}_{aw}(n_a)$) decreases steeply with aggregation number. Conversely, the factor that pertains to aggregate penetration ($\tilde{L}'_{aw}(n_a)$) has an increasing as well as a decreasing component depending on aggregation-number. Factor $\tilde{L}'_{aw}(n_a)$, consequently, changes non-monotonously with increasing aggregation number (see further discussion).

In addition to aggregate flow, trans-barrier gradients also drive a flux of solvent across the barrier

$$J_w = \tilde{L}_w\Delta\mu_{w,i} + \cdots \qquad (2)$$

in proportion to barrier permissiveness to such motion ($\tilde{L}_w$).

Indeed, water flux should be considered in any description of aggregate flow and vice versa. However, such flow only matters when the transportant concentration gradient influences the transport substantially via its effect on the net water activity gradient across the barrier. When such difference is nearly constant eq. 1 alone models the aggregate transport reliably.

Open skin normally contains approximately 75 w% water on the inside and less than 15 w% on the outside [35]. Therefore, water molecules are lost consistently through the skin due to the transepidermal hydration difference; a total of 0.4 μL of water is lost per square centimetre and hour under physiologic conditions [36]. This is insufficient to dehydrate the organ, owing to the practically inexhaustable intra-corporal water reservoir. It also does not suffice to hydrate the skin, due to the quasi-infinite extra-corporal water sink caused by the water evaporation. Water concentration and activity on and below the skin, consequently, is maintained constant under physiologic conditions. To increase permanently the activity of water on the skin one must occlude the organ, or at least reduce the transepidermal water loss, e.g. by raising the ambient humidity close to 100% relative water pressure.

Arguably, the addition of solution or suspension on the skin merely changes quantitatively the balance of material fluxes [17]. The presence of solvated, simple aggregates on the skin increases the volume and concentration of water on the skin, for example. This notwithstanding, the local solvent activity is lower than prior to the addition of simple aggregates, because of the small negative contribution from $\Delta\mu_a/n_a \leq 0$. The same reason can also be given for the diminishment of outwards directed water flux through the organ. Conversely, the evaporative water loss from the skin is increased by similar contribution. The negligibly small transport of simple aggregates through the skin (see further discussion) leads to the final state in which the partly dehydrated aggregates reside on the skin loosing nearly as much water as before the addition of aggregate solution.

Transfersome motion across the skin exemplifies the above mentioned situation: the water activity on either skin side is 'externally controlled' and nearly transportant insensitive, when only aggregate suspension is present on the skin surface.

Barrier resistance is an exponential function of activation energy for the unerlying motion:

$$\mathcal{R}_a = f(r_{pore}, r_a)\exp(G_a^{\#}/RT) \tag{3}$$

$f(r_{pore}, r_a)$ grows monotonously, but typically non-linearly, with increasing pore and transportant radius mismatch ($r_{pore}/r_a \ll 1$). The latter also influences the magnitude of $G_a^{\#}$.

Inverse resistance is identical to barrier penetrability: $\mathcal{P}_a = \mathcal{R}_a^{-1}$. For monomer transport 'penetrability' is called permeability and typically is much higher than for the aggregates. Figure 6 illustrates this schematically. The following rationale and eq. 3 elucidate this dependence more quantitatively.

When a vesicular aggregate of radius r_v is pushed into a narrow pore ($r_{pore} \leq r_v$), the activation energy for the process, in the simplest approximation, is proportional to the relative surface area of a vesicle (r_{pore}^2/r_v^2) [14] and to the membrane elastic energy ($G_{penetration}^{\#} \equiv G^{\#} = \Delta G_{elast}$). The latter is proportional to the inverse square of (relative) vesicle radius and to the membrane elasticity modulus (k_c): $G_{elast} = k_c/2r_v^2 \equiv \kappa/2$ [18].

An aggregate driven into the pore too gently does not protrude deep into the barrier but attains locally stable conformation at the pore entry [37]. Conversely, aggregates that manage to penetrate the pore to a distance comparable to pore radius, owing to their greater adaptability or stronger driving force, profit appreciably from the postulated lower potential energy inside the pore. Such aggregates

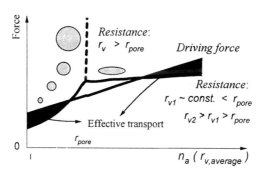

Figure 6: Schematic representation of the forces that affect material transport across a barrier, and their dependence on permeant size or the penetrant aggregation number.

therefore have a propensity to enter the pore spontaneously, but dependent on their deformability. Stress-dependent membrane flexibility maximizes the tendency for pore entry as well as passage.

Stress and composition dependent membrane deformability can be modelled in terms of stress-dependent rigidity factor, $\tilde{\delta}$. This composition dependent [38] factor scales the elastic membrane energy, $G_{elast} = \tilde{\delta}\kappa/2$, relative to that of a standard membrane $G_{elast}(\text{standard}) = \kappa/2$. Small rigidity factor implies small energetic cost of elastic membrane deformation. To obtain highly flexible membranes the requirement $\tilde{\delta} \to 0$ therefore needs to be fulfilled at least under certain stress conditions.

Allowing for anomalous membrane flexibility in eq. 3 predicts barrier resistance in the linear approximation to be inversely proportional to the adjusted membrane elasticity $(\tilde{\delta}\kappa)$ and proportional to relative vesicle size squared (r_v^2/r_{pore}^2) [14]. However, a more general description suggests the individual powers to be different [37]:

$$\mathcal{R} \propto r_v^\alpha / r_{pore}^\beta (\tilde{\delta}\kappa)^\gamma \qquad (4)$$

but typically of the order of 1. Different experimental series, using various driving forces, give $\alpha \sim 1$ and $\beta \leq 1.25$ or $\alpha \sim \beta \sim 1.25$ [17, 39].

The above expression, together with eqs. 1 and 2, approximately describes the transport of water and aggregates across the non-occluded skin. The provision for this is that all boundaries are considered: the first is the actual skin or model skin barrier; the second is a virtual barrier between the volume adjacent to the barrier surface and surrounding air. Since the latter only lets small molecules pass, all phenomenological transport parameters $\tilde{L}_{aw}$, $\tilde{L}'_{aw}$, etc. are here zero with the exception of $\tilde{L}_w$. Moreover, as long as the flux of water through the barrier is smaller than the water loss into the air, we may assume that $\Delta\mu_{w,i}$ remains approximately constant. This implies that eq. 1 can be used with merely one, $\tilde{L}_{aw}$-dependent term. This latter transport parameter is sensitive to the stress-dependent aggregate adaptability.

Hydration gradient. Sensitivity of a composite aggregate to an external activity gradient increases with aggregation number [17]. This becomes clear if we assume, for example, that the activity coefficient of water surrounding a transportable entity gets less; the aggregate then dehydrates partly. Conversely, increasing water activity makes the aggregate hydration more favourable. In either case, the propensity for hydration change depends on the number of hydrophilic entities in an aggregate. The resulting free energy change therefore increases linearly with the aggregate molar hydration energy $(G_{a,hyd0})$, which also is proportional to aggregation number, and with the water activity change. Exceptions are aggregates at and above the limit of solubility: such aggregates do not respond to water concentration increase owing to inter-aggregate interactions. The aggregates therefore shed water below the saturating solvent activity value $(a_w \geq a_{w0})$ and attract, as well as are attracted by, the water molecules in the opposite situation. This is the basis for the observed Transfersome hydrotaxis.

The change in hydration energy arising from the somewhat modified above mentioned system is approximated conveniently by: $\Delta G_{a,hyd}(a_w, n_a) = -G_{a,hyd0}(n_a)$ $\ln(a_w/a_{w0}) \sim n_a G_{a,hyd0}(1) \ln(a_w/a_{w0}) \simeq n_a G_{a,hyd0}(1)\Delta a_w$.

It is necessary that the 'driving energy' exceeds the activation energy for barrier penetration, for significant transport to take place. To a progressive level of approximation this implies that:

$$\Delta G_{a,hyd}(a_w, n_a) > G_a^{\#}$$
$$G_{a,hyd0}(n_a) \ln(a_w/a_{w0}) > r_v^{\alpha}/r_{pore}^{\beta} \, \tilde{\delta}\kappa^{\gamma}$$
$$\Delta a_w \, n_a G_{a,hyd0}(1)(\tilde{\delta}\kappa)^{-\gamma} > r_v^{\alpha}/r_{pore}^{\beta} \qquad (5)$$

the powers α, β and γ being all positive and of order of 1.

The above relationship indicates that lower trans-barrier transport rate, caused by growing transportant size, is compensated by increasing the hydrophilicity and the number of osmotically sensitive molecules in an aggregate exposed to the trans-barrier water activity gradient. This is done by subjecting the aggregates to a higher trans-barrier dehydration pressure but also by making aggregates more deformable and elastic. Deformability therefore remains of paramount importance for the successful transport, especially in the combination with a controllable transport driving gradient increase.

Deformability. Transport, or the lack of it, measured with fairly pure, mono-component phospholipid vesicles (phosphatidylcholine liposomes) provides an estimate for the lower limit of $\mathcal{R}$(liposome). The resistance value for such conventional liposomes exceeds by several orders of magnitude the result determined for Transfersomes. For the highly deformable carriers one accordingly has: $\tilde{\delta}$(Transfersome, $\Delta p \gg 0) \leq 5 \times 10^{-3}$ in the range of high driving gradients. In the other extreme one gets: $\tilde{\delta}$(Transfersome, $\Delta p \to 0) \sim \tilde{\delta}$(liposome) ~ 1. The sigmoidal penetration versus pressure curves pertaining to Transfersomes (cf. figures 1 and 2) are indicative of a rapid, stress-dependent increase in bilayer rigidity factor, $\tilde{\delta}(\Delta p)$.

The conjecture that Transfersome penetration through narrow pores is well regulated and self-optimizing is thus upheld: an excessive initial deformation-stress first decreases the value of $\tilde{\delta}_{transf}(\Delta p)$; this, in turn, reduces the deformation energy and minimizes the stress of penetration; $\tilde{\delta}_{transf}(\Delta p)$ then rises again, which increases the stress, etc. For a low driving pressure such changes are too short-lived to increase the vesicle penetration capability appreciably. In the range of high pressures the resulting high membrane deformability is sufficiently persistent to ensure maximum flow of Transfersomes through the 'confining pores'.

The *local* elastic energy of a Transfersome membrane therefore gets *transiently* low, even lower than RT; this happens when the vesicle rigidity factor is temporarily lowered at some strongly deformed site, $\kappa_{\text{transf}}(\vec{r}) = \bar{\kappa}_{\text{transf}}\tilde{\delta}(\vec{r}) \ll RT \leftrightarrow \tilde{\delta} \ll 1$, due to the stress-induced vesicle membrane adaptation. $\bar{\kappa}_{\text{transf}} \sim RT$ permits relatively easy and fast shifting between the two extremes: while it makes the membrane fluctuations probable (cf. inset to fig. 1) it also maintains vesicle stability [26]. Consequently, at low applied pressures the barrier resistance to the penetration of Transfersomes is typically $> 10^4$ higher than for the pure water, owing to the large aggregate size. Using high pressureschanges the situation: the trans-pore flow of highly deformed Transfersomes now resembles that of water, 1000 times smaller. The sigmoidal pressure-dependence of Transfersome passage through micro-porous membrane (cf. figure 1) and, less directly, the non-linear flow dependence on water activity gradient (cf. figure 2) both support such picture. The composition dependence of the corresponding sigmoidal curve (data not shown) corroborates the conclusion directly.

Data given in figure 2 demonstrate that too small water activity difference causes only an insignificant transport of aggregates across the barrier. Increasing Δa_w enlarges Transfersome flow in a non-linear fashion, indicative of an a_w-dependence of the activation energy and barrier resistance. The measured size dependence ([26] and figure 3) confirms that this activation energy, in the first approximation, is describable with simple elasto-mechanics; the approximately linear relationship between the penetration resistance and the square of (relative) vesicle area ($\mathcal{R} \propto (r_v)^2$) supports such conclusion. The low degree of vesicle fragmentation is supported by the unchanged average vesicle size after the pore passage [26].

High suspension-driving pressure can therefore nearly completely overcome the problem of carrier- and pore-size mismatch (cf. fig. 3). This is true at least for the Transfersomes with a diameter exceeding the pore size by less than the factor of 3 [26].

The observed hypoglycemic effect of Transfersulin appears to be genuine. It is neither mediated by the skin-fluidization nor by the presence of skin lesions or merely the use of particulate carriers. Indeed, none of these factors seems to have played a role in our experiments; the lack of hypoglycemic effect in mice treated with empty carriers, with an insulin solution or with the non-transfersomal insulin suspensions containing an 'improper' blend of similar components on the skin exclude this possibility.

The hypoglycemia measured after non-invasive Transfersulin application, consequently, must be the result of Transfersome-mediated drug transport across the skin. Our studies with radioactive insulin support such a conclusion; so does the similar temporal dependence of C-peptide and glucose concentration changes in the blood after an administration of Transfersulin on the skin [26].

Transfersome components are unlikely to affect the skin barrier directly. Detergents, which are often one of the Transfersome components, for example, are known to affect the permeability barrier of nasal mucosa [5, 30]. If this were the explanation for our results, the measured decrease in the blood glucose concentration after an insulin administration on the skin in Transfersomes would have to increase, or at least remain constant, with increasing detergent concentration. Contrary to such expectation, the epicutaneously applied insulin/mixed micelles mixtures, which contain the highest detergent concentration of all formulations studied, do not cross the animal skin (cf. fig. 1) and have no significant hypoglycemic effect (cf. fig. 5).

Conclusions

The transepidermal water activity gradients can push hydrophilic entities into and across the skin if their resistance to penetration is small enough. Overcoming the size exclusion principle depends on the controlled lowering of the energetic cost for penetrant deformation. In Transfersomes this is realized by combining into a single entity two or more components with sufficiently different packing characteristics and 'spontaneous curvature' to make aggregate membranes highly flexible but not too fragile. Liposomes or mixed lipid micelles fail to fulfill at least one of these requirements and therefore cannot cross the skin.

Drugs administered on the skin in Transfersomes under non-occlusive conditions overcome the barrier efficiently. The carrier-mediated drug delivery across the skin barrier involves repeated passages through the 'virtual channels' in the organ. Such pathways do not exist a priori but can be opened between skin cells and, in particular, between the clusters of such cells by exerting a directed pressure on their junctions. Transfersomes but not liposomes or micelles generate the necessary channels in the skin spontaneously.

The highly deformable, smart carriers, Transfersomes, bring their associated agents through the biological transport barriers, such as the intact skin, reliably and controllably. When loaded with insulin, and applied in a reasonable quantity, Transfersomes induce therapeutically significant hypoglycemia in mice and in humans. Supporting data measured with several other proteins in animals suggest that 'intelligent' materials prepared from the self-optimizing mixtures of lipids offer ample opportunity for non-invasive drug delivery.

Acknowledgements: I have have profitted from, and have enjoyed dealing with, G. Blume, D. Gebauer, A. Paul, H. Richardsen and A. Schätzlein while working on the Transfersome project. Financial supported of IDEA GmbH and of Deutsche Forschungsgemeinschaft (through the research grant Ce 9/1) is gratefully acknowledged.

References

1. Saffran, M.; Sudesh, G.K.; Savariar, C.; Burnham, J.C. Williams, F.; Neckers, D.C. *Science* **1986**, *233*, 1081-1084.
2. Das, N.; Basu. M. K.; Das, M. K. *Biochem. Int.* **1988**, *16*, 983-989.
3. Cho, Y. W.; Flynn, M. *Lancet* **1989**, *23-30*, 1518-15.
4. Liedtke, R.K.; Suwelack, K.; Karzel, K. *Arzneimittelforschung* **1990**, *40*, 880-883.
5. Aungst, B. J.; Rogers, N. J.; Shefter, E. *J. Pharm. Exp. Ther.* **1988**, *244*, 23-27.
6. Chiou, G. C. Y.; Chuang, C. Y.; Chang, M. A. *J. Ocul. Pharmacol.* **1989**, *5*, 81-91.
7. Kimmerle, R.; Griffing, G.; McCall, A.; Ruderman, N. B.; Stoltz, E.; Melby, J. C. *Diab. Res. Clin. Pract.* **1991**, *13*, 69-76.
8. Ogiso, T.; Nishioka, S.; Iwaki, M. *Biol.; Pharm.; Bull,* **1996**, *8*, 1049-54.

86

9. Christophers, E.; Schubert, M.; Goos, C. In *Pharmacology of the skin*; Editors, Greaves, M.W.; Shuster, S. Springer–Verlag, Berlin, 1989, Volume 1; pp 3–30.
10. Fartasch, M.; Bassukas, I. D.; Diepgen, T. L. *Br. J. Dermatol.* **1993**, *128*, 1-9.
11. Schätzlein, A. Cevc, G. *Br. J. Dermatol.* **1998**, in the press.
12. *Transdermal Drug Delivery. Developmental Issues and Research Initiatives.* Hadgraft, J.; Guy, R. H. Eds.; Marcel Dekker, New York, **1989**.
13. Potts, R. O.; Guy, R. H. *Pharm. Res.* **1992**, *9*, 663-669.
14. Cevc, G. In *Handbook of Physics of Biological Systems*; Editor, R. Lipowsky; Elsevier Science, Amsterdam, 1995 Vol. I; Chapter 9, 441-466.
15. Cevc, G.; Blume, G. *Biochim. Biophys. Acta* **1992**, *1104*, 226-232.
16. Cevc, G.; Blume, G.; Schätzlein, A.; Gebauer, D.; Paul, A. *Adv. Drug Del. Rev.* **1996**, *418*, 349–378.
17. Cevc, G. *Crit. Rev. Therap. Drug Carrier Syst.* **1996**, *13*, 257-388.
18. Helfrich, W. *Z. Naturforsch.* **1973**, *28c*, 693-703.
19. Needham, D.; Evans, E. *Biochemistry* **1988**, *27*, 8261–8269.
20. Lipowsky, R. *Nature* **1991**, *349*, 475-481.
21. Mayer, L. D.; Hope, M. J.; Cullis, P. R. *Biochim. Biophys. Acta* **1986**, *856*, 161-168.
22. Zellmer, S.; Pfeil, W.; Lasch, J. *Biochim. Biophys. Acta* **1995**, *1237*, 176–182.
23. Schreier, H.; Bouwstra, J. *J. Contr. Rel.* **1994**, *30*, 1-15.
24. Lichtenberg, D.; Robson, R. J.; Dennis, E. A. *Biochim. Biophys. Acta* **1983**, *737*, 285-304.
25. Cevc, G. Eur. pat. 91 114 163.8-2114, **1991**.
26. Cevc, G.; Gebauer, D.; Schätzlein, A.; Blume, G.; Stieber, J. *Biochim. Biophys. Acta* **1998**, *1368*, 201-215.
27. Cevc, G. *Exp. Opinion Invest. Drugs* **1997**, *6*, 1887-1937.
28. Cevc, G.; Blume. G.; Schätzlein, A. *J. Contr. Rel.* **1996**, *45*, 211–226.
29. Cevc, G.; Schätzlein, A.; Blume, G. *J. Contr. Rel.* **1995**, *36*, 3-16.
30. O'Hagan, D. T.; Illum, L. *Crit. Rev. Therap. Drug Carrier Syst.* **1990**, *7*, 35-97.
31. Wearley, L. L. *Crit. Rev. Therap. Drug Carrier Syst.* **1991**, *8*, 331-394.
32. Paul, A.; Cevc, G.; Bachhawat, B. K. *Eur. J. Immunol.* **1995**, *25*, 35221-3524.
33. Paul, A.; Cevc, G. *Vaccine Res.* **1995**, *4*, 145–164.
34. Cevc, G. In *Frontiers in Insulin Pharmacology*; Editors, M. Berger & F. A. Gries; Georg Thieme Verlag, Stuttgart, 1993; 161–169.
35. Warner, R.R.; Myers, M.C.; Taylor, D.T. *J. Invest. Dermatol.* **1988**, *90*, 218–224.
36. Potts, R. O.; Freancoeur, M. L. *Proc. Natl. Acad. Sci. (USA)* **1990**, *87*, 3871-3873.
37. Gompper, G.; Kroll, D.M. *Phys. Rev. E* **1995**, *52*, 4198-4208.
38. Leibler, S. *J. Physique (France)* **1986**, *47*, 507-516.
39. Gebauer, D. **1998**, Ph. D. Thesis, The Technical University of Munich.

Chapter 7

pH Oscillation of a Drug for Temporal Delivery

Steven A. Giannos and Steven M. Dinh

Novartis Pharmaceuticals Corp., 25 Old Mill Road,
P.O. Box 100, Suffern, NY 10901

The growing interest in chronopharmacology has led to the need for temporally controlled or modulated delivery systems to optimize drug therapies. A variety of physical and chemical concepts are being explored to develop innovative technologies to respond to this need. Ultrasound, magnetism and electrical principles have been introduced as external applicators which modulate the delivery profile. Advances in non-linear chemical kinetics open the gateway for chemical oscillators to be used as internal timing mechanisms for drug delivery or other devices. The coupling of a chemical oscillator with membrane diffusion offers a unique opportunity to design novel, temporally controlled, drug delivery systems which are internally regulated.

Recent advances in chronopharmacology demonstrate the importance of biological rhythms to the dosing of medications (1-5). Circadian rhythms, seasonal cycles and fertility cycles should be considered as part of optimizing the therapeutic effect of a drug and its delivery system. Studies indicate that the onset of diseases, such as specific cancers, show strong seasonal dependency. Certain diseases, asthma and angina for example, occur more frequently at particular times of the day; in the afternoon or evening for asthma and in the morning hours for angina (6). The secretion of gastric acid has been found to increase in the afternoon, leading physicians to recommend that ulcer medications be taken in the afternoon for better efficacy. Moreover, Parkinson's disease (7-8), which is treated by L-Dopa and angina (9-13) which is treated by nitroglycerin would be better managed with chronobiological therapy. Additionally, tolerance to various drugs and skin irritation and sensitization caused by medications may require intervals during which no drug is administered.

These examples challenge the early hypothesis of zero-order release as being the optimal method of administering medications, and open opportunities for technological innovation. Therefore, the need for temporally controlled drug delivery systems to optimize therapy has initiated the evaluation of pulsatile delivery. The development of pulsatile delivery systems can be categorized as either externally modulated or self-regulated (14). Externally modulated drug delivery is accomplished by the response of a device to an externally generated signal. Such methods include ultrasound (sonophoresis), magnetism and electric potential (iontophoresis). Most of these methods are at the experimental stage (15). Self-regulated delivery systems generally consist of a rate-control mechanism and a functionalized polymer or hydrogel matrix (16-17). The modulation of drug delivery through internally regulated delivery systems has only recently been considered. For instance, interest in the development of membranes with oscillating permeability for periodic drug delivery, inspired by the Belousov-Zhabotinskii reaction, has been initiated (18).

Through the manipulation of oscillating reactions, chemical oscillators may be used as the internal timing mechanism in temporal drug delivery or other devices. Drug efficacy could be improved and tolerance or sensitization may be minimized by providing drug-free periods (19).

Temporally controlled transdermal drug delivery

The simplest type of chronotherapy is a biphasic profile, in which the drug concentration changes from a high to a low level (or vice versa) over time. Although the system can be physically applied or removed to alter the drug level, patient compliance to this procedure may be difficult, particularly during inconvenient hours. To generate a biphasic profile, the delivery system may utilize an external regulator, or it can use passive diffusional means to create an abrupt change in drug release. The same mass-transfer principles are applied to the design of a biphasic delivery system, by examining the factors that control the transport surface area, the mass-transfer resistance and the driving force (20). In practice, altering the transport area in transdermal delivery has not been pursued because of difficulties to control this process. The strategy of changing the mass-transfer resistance can be applied primarily to the components of the delivery system. For instance, Bae et al. developed an IPN (interpenetrating polymer network) membrane wherein a bimodal release rate (two pulses per dose) was achieved (21). This approach, however, is limited to drugs that are highly permeable across skin, such as nicotine, and uses a membrane-controlled design to modulate drug delivery. The third strategy is to modulate the driving force of the drug by controlling the change in the solubility of the drug in skin in the presence of a permeation enhancer. In turn, the instantaneous concentration of the permeation enhancer is regulated by the control membrane of the delivery system, or by depleting the enhancer from the system. The application of this principle to produce a biphasic delivery profile for lipophilic drugs, such as nitroglycerin, was demonstrated by Dinh et al. (20) using ethanol as a permeation enhancer.

Application of chemical oscillators. A novel technology, using chemical oscillators, is being developed to generate a periodic release of a drug or active ingredient without external power sources and/or electronic controllers. The strategy is based on the observation that a drug may be rendered charged or uncharged relative to its pK_a value. Since only the uncharged form of a drug can permeate across lipophilic membranes, including the skin, a periodic delivery profile may be obtained by oscillating the pH of the drug solution (*19*). The challenge centers on developing a method to change the pH without the need of a continuous external modulator.

The approach that we have undertaken originates from the nonlinear kinetics of chemical oscillators. Chemical oscillating reactions have been known for nearly a century, with the Belousov-Zhabotinskii (BZ) and the Landoldt reactions being the two best characterized oscillators. The BZ systems, which are models for studying a wide variety of temporal and spatial instabilities in chemical systems (*22*), are generally accepted as the metal-ion-catalyzed oxidation and bromination of an organic substrate by acidic bromate. These colorful reactions illustrate the principles of chemical oscillators through cyclic changes in the color of the solutions.

The usual practice for the study of oscillating systems has been to use closed (batch) reactors or an open system (continuous flow stirred tank reactors (CSTR)). The recent description of using a "semibatch reactor" as an additional tool, is an appealing and simple intermediate method to study pH oscillating systems (*23*).

The pH oscillators consist of those oscillating chemical reactions in which there is a large amplitude change in the pH, and in which the pH change is an important driving force rather than a consequence or an indicator of the oscillation (*24-26*). The pH of a solution can be oscillated over a range of pH values from 2 to 10 by the reduction and oxidation (redox) reactions of salts, such as permanganates, iodates, sulfates, chlorates, or bromates. The first pH oscillator, the hydrogen peroxide-sulfide reaction, was discovered only fifteen years ago. Approximately 14 pH oscillators, listed in Table 1, have been identified (*26*). The mixed Landoldt reaction (iodate-thiosulfate-sulfite), that oscillates the pH between 6.5 and 4.0, has been studied extensively (*27-28*). With this oscillating system there is a characteristic "spike" where the pH minimum is just above 4.0. If the pH falls below 4.0 for any length of time, the iodate-iodide (Dushman) reaction predominates and the solution turns brown. A proposed mechanism for this reaction is as follows (*28*):

$$A + B \rightarrow Y \tag{1a}$$

$$A + B + X \rightarrow P_1 \tag{1b}$$

$$A + Y + X \rightarrow 2X + P_2 \tag{1c}$$

where A corresponds to iodate, B to thiosulfate, Y to hydrogen sulfite, X to hydrogen ion, P_1 to tetrathionate and P_2 to sulfate. The basis for the oscillatory behavior is the *alternation* of the autocatalysis of sulfite and the consumption of hydrogen ion and

90

Table I.   pH Regulated Oscillating reactions (ref. *24, 25*)

	SYSTEM	pH
1.	IODATE-SULFITE-THIOUREA	3.5-7.0
2.	IODATE-SULFITE-THIOSULFATE	4.1-6.5
3.	IODATE-SULFITE-FERROCYANIDE	2.5-8.0
4.	IODATE-HYDROXYLAMINE	2.8-5.5
5.	PERIODATE-HYDROXYLAMINE	
6.	PERIODATE-THIOSULFATE	4.0-6.0
7.	PERIODATE-MANGANESE(II)	3.5-4.5
8.	HYDROGEN PEROXIDE-FERROCYANIDE	5.0-7.0
9.	HYDROGEN PEROXIDE-THIOSULFATE-COPPER(II)	6.0-8.0
10.	HYDROGEN PEROXIDE-BISULFITE-THIOSULFATE	
11.	PEROXODISULFATE-THIOSULFATE-COPPER(II)	2.3-3.0
12.	BROMITE-IODIDE	
13.	BROMATE-SULFITE-FERROCYANIDE	4.5-6.5
14.	BROMATE-SULFITE-THIOSULFATE	

SOURCE: Adapted from ref. 30.

the formation of sulfite (27). The authors have discovered that poly(2-acrylamido-2-methyl-1-propane-sulfonic acid) (PAMPS), as well as other sulfonic acid polymers, can be substituted for sulfuric acid in the modified Landoldt oscillator (29).

To integrate pH oscillators into the design of a drug delivery system, the underlying characteristic times need to be analyzed. The key parameter for system design is the ratio of the characteristic time for permeation to the characteristic time of the oscillation of the driving force so that the time of drug permeation is shorter than the oscillation period. This ratio must be small to produce a temporally controlled delivery profile. As an example, a periodic drug release profile can be obtained by ensuring that the period of oscillation in the drug input, through the use of a pH chemical oscillator, is longer than the permeation of the drug across all diffusional barriers. In practice, the period of the pH oscillator must be longer than the pharmacokinetic time lag, which is related to the sum of the characteristic times for obtaining steady state flux through the membranes and steady-state plasma levels. The following analysis illustrates how this key parameter controls the delivery of a drug across a membrane.

Consider an ideal situation where a drug with a known pK_a is in an infinite reservoir, in which the pH of the solution is periodically changed by a pH chemical oscillator. The instantaneous concentration of the uncharged form of the drug [$C(t)$], which can permeate across a hydrophobic membrane, is given by:

$$C(t) = C_{max} \sin (\omega t)\, H(t) \tag{2}$$

where C_{max} is the maximum concentration, t is time, and ω is the frequency (or $2\pi/\omega$ is the period of oscillation) and $H(t)$ is the Heaviside function. The frequency, ω, is controlled by the kinetics of the pH oscillation, and hence by the selection of the chemical oscillator. The flux of the drug across the membrane, expressed in a dimensionless form, is then given by:

$$-\frac{l j(t, x=0)}{2KDC_{max}} = \tag{3}$$

dimensionless flux

$$\sum_{n=1}^{\infty} (-1)^{n+1} n^2 \left\{ \frac{1}{\left(n^4 + \lambda^2\omega^2\right)^{1/2}} \sin\left[\omega t - \tan^{-1}\left(\frac{\lambda\omega}{n^2}\right)\right] + \frac{\lambda\omega}{\left(n^4 + \lambda^2\omega^2\right)} e^{-n^2 t/\lambda} \right\}$$

forced oscillation   *diffusion*

where l, K, D and λ $(= l^2/\pi^2 D)$ are the thickness of the membrane, the partition coefficient, the diffusivity and the characteristic time of permeation. The characteristic time for permeation, such as the lag time, is governed by the diffusivity and the thickness of the membrane. The first term in the right-hand-side of the above equation describes the contribution to the flux by the imposed periodic change of the driving force as modulated by the frequency-filtering effect of the membrane. The second term represents the transient permeation. Consequently, if the conditions are set up such that the second term dominates ($\lambda\omega \gg 1$), then the output flux decays to zero, the mean driving flux force for diffusion. However, if the conditions are set up such that the first term dominates ($\lambda\omega \ll 1$), then the flux of the drug across the membrane oscillates with a frequency distribution reflecting the oscillation of the driving force of the permeant and the filtering effect of the membrane. Defining the set of conditions, in which $\lambda\omega < 1$, is the underlying principle for the development of the temporally controlled delivery system (*30*). This limit is exemplified by a single oscillation of the fluxes of benzoic acid and nicotine across an ethylene vinyl acetate copolymer membrane containing 28% vinyl acetate.

Experimental Procedure

Details of the experimental procedures and results have been described previously (*19*). Studies of pH oscillators were performed in a simple semibatch reactor or a custom-made diffusion cell apparatus with a 10 cm^2 area. The reactants were pumped into the reaction/donor cell using a peristaltic pump and 1.14 mm i.d. Tygon pump tubing.

Determination of Permeated Benzoate Ion at Constant pH Values. A modified assay method was developed for the determination of benzoic acid permeation across EVA membranes at constant pH values. The permeation of a model compound (benzoic acid) across a model membrane (28% EVA, 2 mil thick, 32°C) was characterized at discrete pH values (*19*).

Effect of reversal of order of adding reactants. Experiments were performed using semibatch conditions with a 2 mil, 28% EVA film at 32°C and a 75 mL donor cell. Sodium iodate (solution A) was introduced into 75 mL of Solution B containing sodium sulfite, sodium thiosulfate, sulfuric acid and sodium benzoate at a rate of 0.080 mL/min. using a peristaltic pump. The reversal of the order of addition resulted in a single period. The presence of 100% of the model acid drug in solution B, allowed it to respond to the changes in the solution pH (*19*).

The nicotine experiment was run using the same semibatch conditions with a 2 mil, 28% EVA film at 32°C and a 75 mL donor cell. Sodium iodate (solution A) was introduced into 75 mL of Solution B containing sodium sulfite, sodium thiosulfate, sulfuric acid and 0.75 mL of nicotine free base at a rate of 0.080 mL/min. using a peristaltic pump. In this experiment the donor solution was allowed to stand for 60 minutes in order for the nicotine to passively diffuse across the membrane. After this time period, the iodate solution was added and the oscillation reaction begun. The presence of 100% of the model basic drug in solution B, allowed it to react to the changes in the solution pH (19).

Results and Discussion

The permeation of benzoic acid across the membrane (Figure 1) responded to the change in pH during the addition of the iodate solution; increasing in flux when the pH of the solution decreased toward the pK_a of benzoic acid (pK_a of benzoic acid being 4.2) and decreasing in flux when the pH of the solution increased. The diffusional lag time of benzoic acid was found to be approximately 20 minutes.

The permeation of nicotine across this ethylene vinyl acetate copolymer membrane is shown in Figure 2. The pK_a values of nicotine are 3.4 and 7.9. In this illustration, nicotine free base permeated across the membrane during the first 60 minutes, in the absence of the iodate oscillator. The diffusional lag time of nicotine across this membrane was approximately 30 minutes. Thereafter, the pH oscillatory reaction was initiated by the addition of the iodate solution. As shown in Figure 2, the flux of nicotine reached a maximum at approximately 90 minutes, then it decreased in response to the lowering of the pH of the solution. The flux of nicotine increased again, responding to the increasing pH of the oscillator reaction (19).

One limitation for the formulation of oscillator-diffusion systems, described earlier in equation (3) as $\lambda\omega < 1$, has been the need for a pH oscillating reaction that has a long enough periodic time, and a longer low pH state within a cycle, to allow diffusion across a membrane. Recently, Rabai and Hanazaki described the bromate oscillator system, using marble as the acid accepting component (31). The basis for the oscillatory behavior is the autocatalysis of sulfite and the consumption of hydrogen ion by marble. The kinetics of the pH oscillations using bromate as the oxidizer and marble chips as the hydrogen acceptor are slower, thereby lengthening the oscillation periods to approximately an hour, as opposed to seconds in the modified Landoldt reaction (31). We were able to obtain reproducible oscillations when using the conditions described by Rabai and Hanazaki (Figure 3). Nicotine hemi-sulfate salt was found to be compatible when added to the bromate solution. Successful pH oscillations (Figure 4), with period lengths of approximately an hour with longer low pH states, were obtained, suggesting that the bromate-marble pH oscillating reaction may be useful in the development of temporally controlled drug delivery systems. Additionally, the bromate-marble pH oscillating reaction is an attractive step towards the development and adaptation of chemical oscillating reactions for use in medical products.

The current limitations for the formulation of these coupled oscillator-diffusion systems are:

1. The ionized and unionized forms of the drug have a finite buffer capacity that results in damping of the pH oscillations. A concentration must be selected as a compromise between an acceptable amount of damping and the desired drug flux.

94

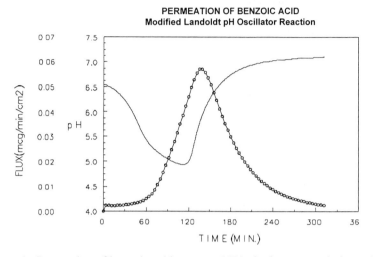

Figure 1. Permeation of benzoic acid across a 28% vinyl acetate, ethylene vinyl acetate copolymer membrane. The flux of benzoic acid [circle line] changes relative to the change in the pH of the solution [solid line]. (Reproduced with permission from ref. 30. Copyright 1996 Gordon and Breach).

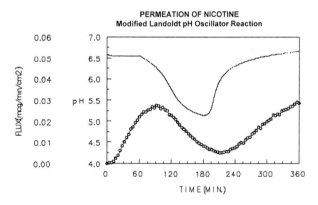

Figure 2. Permeation of nicotine across a 28% vinyl acetate, ethylene vinyl acetate copolymer membrane. The flux of nicotine [circle line] changes relative to the change in the pH of the solution [solid line]. (Reproduced with permission from ref. 30. Copyright 1996 Gordon and Breach).

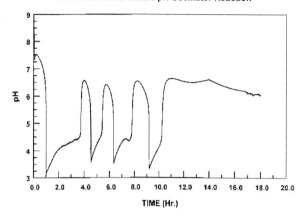

Figure 3. pH oscillations obtained from the Bromate-Sulfite-Marble semibatch reaction.

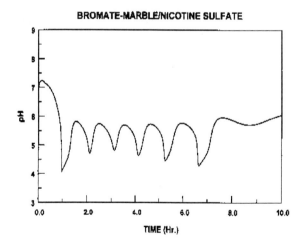

Figure 4. pH oscillations obtained from the Bromate-Sulfite-Marble semibatch reaction containing nicotine hemi-sulfate.

2. The oscillations are extremely sensitive to initial conditions. A regime where oscillations occur and which is relatively insensitive to these conditions must be determined empirically.
3. The desired frequency and amplitude are not always possible for a given oscillator. The time lags of 20 and 30 minutes for benzoic acid and nicotine, respectively, necessitate the use of a lower frequency than is typically observed with the iodate-thiosulfate oscillator. Membranes with time lags on the order of 5 minutes would exhibit multiple oscillations. Mathematical modeling and discovery of new oscillators should increase the predictability and the adaptability of this technology (*30*).

The implementation of this concept to a transdermal delivery system is ongoing. As an example, a user activated system may be employed, where the reactants are initially separated in different compartments. The user-activated system is a two chamber reservoir system, where the compartments are separated by a weak seal during storage. The drug can be in either compartment, provided that the drug is stable in that environment. The user activates the system, prior to usage, by breaking the weak seal. The contents from the two compartments are mixed to form the pH-oscillating solution. The uncharged state of the drug then permeates across lipophilic barriers, such as the control membrane of the delivery system and skin. The use of a lipophilic control membrane ensures that the reactants, which are charged species, do not permeate out of the drug delivery system to elicit adverse biological responses. Beyond this preliminary investigation, other strategies are being considered, such as reacting a drug with chemical oscillators to alter its charge state, or altering the charge state of the control membrane or a surface layer of the control membrane by oscillatory reactions. Numerous concepts have yet to be examined in this fertile field where technology innovation can be used to improve the quality of life by tackling unmet medical needs.

Literature Cited

1. Smolensky, M. H and Labrecque, G. *Pharmaceutical News* **1997**, *4*, *No. 2*.
2. Hrushesky, W. *J. Cont. Rel.* **1992**, *19*, 363.
3. Hrushesky, W. *THE SCIENCES* **1994**, July/Aug., 33-37.
4. Lemmer, B., and Bruguerolle, B. *Clinical Pharmacokinetics* **1994**, *26*, 6, 419-427.
5. Redfern, P., Minors, D., and Waterhouse, J. *Chronobiology International* **1994**, *11*, 4, 253-265.
6. Lemmer, B. *Adv. Drug Del. Rev.* **1991**, *6*, 83-100.
7. Nutt, J. G. In *Neurological Applications of Implanted Drug Pumps;* Penn, R. D. Ed.; The New York Academy of Sciences, New York NY, **1988**, 194-199.
8. Altar, C. A., Berner, B., Beall, P., Carlsen, S. F., and Boyar, W. C. *Mol. Pharmacol.* **1988**, *33*, 690.

9. Yates, F. E., and Benton, L. A. In *Temporal Control Drug Delivery*, Hrushesky, J. M., Langer, R. S. and Theewes, F. Eds.; The New York Academy of Sciences, New York NY, **1991**, 38-56.
10. Wolff, H. M., and Bonn R. *Eur. Heart J.* **1989**, *10*, 26-29.
11. Fujimura, A., Ohashi, K.-I., Sugimoto, K., Kumagai, Y., and Ebihara, A. *J. Clin. Pharmacol.* **1989**, *29*, 909-915.
12. Krepp, H. P. and Turpe, F. *Eur. Heart J.* **1989**, *10*, 36-42.
13. Wiegand, A., Bohn, R., Trenk, D., and Jahnchen, E. *J. Clin. Pharmacol.* **1992**, *32*, 77-84.
14. Kost, J. and Langer, R. *Adv. Drug Del. Rev.* **1991**, *6*, 19-50.
15. Kost, J. and Langer, R. *Trends Biotechnol.* **1992**, *10*, 127-131.
16. Heller, J. *Critical Reviews in Therapeutic Drug Carrier Systems*, **1993**, *10*, 3, 253-305.
17. Yoshida, R., Sakai, K., Okano, T., and Sakurai, Y. *Adv. Drug Del. Rev.* **1993**, *11* 85-108.
18. Siegel, R. A. and Pitt, C. G. *J. Cont. Rel.* **1995**, *33*, 173-188.
19. Giannos, S. A., Dinh, S. M., and Berner, B. *J. Pharm. Sci.* **1995**, *84*, 5, 539-543.
20. Dinh, S. M., Comfort, A. R., Shevchuk, I., and Berner, B. *J. Memb. Sci.* **1994**, *89*, 245-255.
21. Bae, Y. H., Okano, T., Ebert, C., Heiber, S., Dave, S., and Kim, S. W. *J. Cont. Rel.* **1991**, *16*, 189-196.
22. Zhabotinskii, A. M. In *Oscillations and Traveling Waves in Chemical Systems*; Field, R. J., Burger, M., Eds.; Wiley-Interscience: New York NY, **1983**.
23. Rabai, G. and Epstein, I. R. *J. Am. Chem. Soc.,* **1992**, *114* 1529-1530.
24. Rabai, G. Orban, M. and Epstein, I. R. *Acc. Chem. Res.,* **1990**, *23* 258-263.
25. Luo, Y. and Epstein, I. R. *J. Am. Chem. Soc.,* **1991**, *113* 1518-1522.
26. Epstein, I. R. and Orban, M. in *Oscillations and Traveling Waves in Chemical Systems*; Field, R. J., Burger, M., Eds.; Wiley-Interscience: New York NY, **1983**.
27. Rabai, G. and Beck, M. T. *J. Phys. Chem.* **1988**, *92*, 4831-4835.
28. Rabai, G. and Beck, M. T. *J. Phys. Chem.* **1988**, *92*, 2804-2807.
29. Giannos, S. A., Dinh, S. M. and Berner, B. *Macromol. Rapid Comm.* **1995**, *16*, 527-531.
30. Giannos, S. A. and Dinh, S. M. *POLYMERNEWS* **1996**, *21*, 118-124.
31. Rabai, G. and Hanazaki, I. *J. Phys. Chem.* **1996**, *100*, 10615 10619.

Chapter 8

pH-Hysteresis of Glucose Permeability in Acid-Doped LCST Hydrogels: A Basis for Pulsatile Oscillatory Drug Release

J.-C. Leroux[1] and R. A. Siegel

Departments of Biopharmaceutical Sciences and Pharmaceutical Chemistry, University of California, San Francisco, CA 94143–0446

Some drugs and hormones are efficacious only when delivered in a periodic, pulsatile manner. This observation motivates the construction of periodic pulsatile delivery devices. We report progress towards an implantable device that pulsates autonomously without prompting from an external energy source. The device is conceived by combining understanding of oscillating chemical reactions and of the properties of gels which undergo first-order swelling phase transitions. Chemical oscillators typically include a reaction whose rate law shows hysteresis, coupled to a feedback reaction which causes the system to switch back and forth between branches of the former reaction. First order swelling transitions in hydrophobic acidic gels also feature hysteresis with respect to pH. Coupling swelling hysteresis to transport of the substrate and product of an enzymatic reaction, we predict periodic, pulsatile behavior of the gel/enzyme system. Here we summarize the theoretical basis of the device, demonstrate hysteresis in permeability of a selected membrane, and present initial evidence for oscillatory device behavior.

In the past decade, there has been a growing interest in the use of hydrogels based on polymers presenting a lower critical solution temperature (LCST) for the controlled delivery of drugs (*1*). Among them, crosslinked poly(*N*-isopropylacrylamide) (NIPA) gels have been extensively characterized (*2-4*) and studied for various pharmaceutical applications (*5-7*). These hydrogels demonstrate a sharp volume transition near 31-35°C (*3,8,9*), going from a highly swollen state at low temperatures to a collapsed dehydrated state at higher temperatures. The phase transition has been identified with an increase in conformational entropy of water as it leaves the gel, which compensates for the decrease in mixing and elastic entropies as the relatively hydrophobic network collapses near the phase transition point. The overall contribution to the free energy of the gel system is negative leading to this spontaneous and reversible behavior (*10*).

[1]**Current Address:** Faculty of Pharmacy, University of Montreal, Quebec H3C 3J7, Canada.

NIPA hydrogels have been shown to achieve on-off regulation of the release of low molecular weight drugs such as indomethacin (11).

Under certain conditions, the phase transition of NIPA hydrogels can also be induced by other stimuli such as organic solvents (12), light (13) or salts (14), but an interesting prospect is the possibility of producing hydrogels responding to pH. The introduction of small amounts of ionizable comonomers such as sodium acrylate (SA) (3,8) or (diethylamino)-ethyl methacrylate (15) into crosslinked NIPA gels increases the volume changes associated with gel transitions and can elicit a discontinuous swelling transition. Depending on the amount of comonomer added and pH, these gels may exist in a swollen state at 37°C because of the repulsion of ionic groups which maintains the network in an expanded form. In this case, a change in pH affects the ionization state of the comonomer and hence the swelling of the gel. Dong and Hoffman (16) synthesized hydrogels composed of NIPA, acrylic acid and vinyl terminated polydimethylsiloxane for the enteric delivery of indomethacin. At low gastric pH the gel is in a collapsed state because its critical temperature is below 37°C, and the drug remains entrapped in the polymeric network. At pH 7.4 the acrylic acid becomes ionized and the gel's critical temperature shifts above 37°C, leading to swelling of the gel and subsequent release of indomethacin. Similar systems have been reported for the controlled release of macromolecular compounds (6,17,18).

One of the most interesting pharmaceutical applications of these pH-sensitive hydrogels concerns the pulsatile delivery of drugs. Pulsatile drug release is suitable for the administration of drugs of tolerance such as nitroglycerin (19) and for certain hormone replacement therapies where it has been found that periodic fluctuations in the endogenous release pattern are strongly correlated with hormonal effect (20,21). Several open loop drug delivery devices that can deliver a compound in bursts at distinct periods without external activation have been described (22-24). However, many of these systems can only provide a finite number of pulses. More recently, the possibility of coupling a chemical oscillator with release of a drug has been investigated. (25,26). By using the Landolt mixed oscillator, a well-known pH-oscillating reaction (27), Giannos and coworkers (25) were able to fluctuate the ionization state of nicotine or benzoic acid and hence modulate the permeation of these drugs across a lipophilic membrane. Further developments of this system will require discovery of new chemical oscillators showing a sufficiently long period of oscillation to be efficiently coupled with the slow diffusion of the drug through the membrane (28). Similarly, it has been demonstrated that a pH-oscillating chemical reaction can be coupled to a pH-sensitive hydrogel to produce oscillations of the gel swelling and deswelling (29,30). As illustrated in figure 1, the change in the swelling state of the gel as a function of pH could be used as on-off regulator for the release of compounds. Although the schemes described in this paragraph illustrate the potential of chemical oscillators in the design of pulsatile drug delivery systems, some limitations remain. First, the oscillations are maintained only if the system is continuously replenished with fresh reactants. Second, certain chemicals used in most chemical oscillators are toxic reactants which are not welcome in drug delivery applications, especially if the devices are to be implanted.

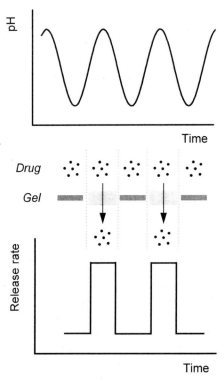

Figure 1. General scheme of oscillating gels for pulsatile, oscillatory drug release.

In our laboratory we are attempting to construct an implantable drug delivery device that pulsates autonomously without prompting from an external energy source. The device functions by dissipating the chemical energy supplied by a constant concentration of endogenous glucose. Glucose exists in abundance in the body at a reasonably well controlled concentration, at least in nondiabetic individuals. The system is conceived by combining an understanding of mechanisms underlying oscillating chemical reactions, and of the properties of pH-sensitive gels which undergo first-order swelling phase transitions. Chemical oscillators typically include a reaction whose rate law shows hysteresis coupled to a feedback reaction which causes the system to switch back and forth between branches of the former reaction (*31,32*). As we will see below, first order swelling transitions in hydrophobic acidic gels also feature hysteresis with respect to pH. Coupling swelling hysteresis to transport of the reactant (glucose) and product (H^+) of an enzymatically (glucose oxidase) driven reaction can theoretically lead to periodic, pulsatile behavior in suitably constructed gel/enzyme systems (*33,34*). In this contribution we summarize recent experimental results regarding hysteresis in a particular hydrogel membrane with respect to proton concentration and provide preliminary evidence for oscillation.

Theoretical section

Figure 2 illustrates the principle behind the proposed device. A drug-containing chamber is separated from the external environment by a gel membrane, which swells with increasing pH and collapses with decreasing pH. Outside the device is a constant concentration of glucose, which is a substrate (S) for enzymes (enz) located inside the chamber. Glucose (S) diffuses into the chamber through the membrane and reacts to form hydrogen ion product (P) which causes the membrane collapse and subsequent shutdown of glucose and drug permeation. Eventually, H^+ concentration inside the device depletes by diffusion through the membrane, leading to reswelling of the membrane and restoration of permeability to glucose and drug. The H^+ concentration inside the chamber inhibitorily affects glucose permeability, and may also affect H^+ permeability. The membrane is assumed to switch between a high permeability state, H, and a low permeability state, L, with hysteresis. Let

A = membrane area
V = chamber volume
C_S = substrate (glucose) concentration inside the device
C_P = product (H^+) concentration inside the device
C_S^* = external glucose concentration
k_{enz} = first order rate constant for enzyme reaction (ignore saturability)
$K_{S,H}$ = membrane permeability to glucose in the H state
$K_{S,L}$ = membrane permeability to glucose in the L state
$K_{P,H}$ = membrane permeability to H^+ in the H state
$K_{P,L}$ = membrane permeability to H^+ in the L state
$C_{P,H \to L}$ = H^+ concentration where H→L transition occurs
$C_{P,L \to H}$ = H^+ concentration where L→H transition occurs

When the membrane is in the H state, the governing equations are

$$\frac{dC_S}{dt} = (A/V)\, K_{S,H}\, (C_S^* - C_S) - k_{enz}\, C_S \,; \quad \frac{dC_P}{dt} = k_{enz}\, C_S - (A/V)\, K_{P,H}\, C_P \qquad (1)$$

The membrane switches from the H to the L state when C_P increases past $C_{P,H \to L}$. In the L state the governing equations are:

$$\frac{dC_S}{dt} = (A/V)\, K_{S,L}\, (C_S^* - C_S) - k_{enz}\, C_S \,; \quad \frac{dC_P}{dt} = k_{enz}\, C_S - (A/V)\, K_{P,L}\, C_P \qquad (2)$$

The membrane switches to the H state when C_P drops below $C_{P,L \to H}$. Equations 1 and 2 are similar to those presented previously (36) except for the hysteretic behaviors of K_S and K_P, and the absence of need to assume a finite relaxation time for the membrane permeabilities in response to changes in C_P.

It is possible using techniques of bifurcation theory (*35*) to determine conditions under which the system will oscillate between the H and L states. When the enzyme reaction is not rate limiting, it can be shown (Zou and Siegel, submitted) that the solutions of equations 1 or 2 are periodic when the following condition is satisfied:

$$\frac{K_{P,H}}{K_{S,H}} \, C_{P,H \to L} < C_S^* < \frac{K_{P,L}}{K_{S,L}} \, C_{P,L \to H} \tag{3}$$

These oscillations are expected to drive periodic, pulsatile delivery of an active agent if that agent's permeability is dependent on membrane state. We shall not deal at this point with delivery of such drugs, since experiments to date have not included them.

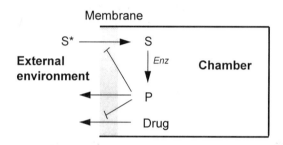

Figure 2. General scheme for proposed drug delivery oscillator.

The enzymatic reaction which catalyses the conversion of glucose to hydrogen ion is depicted in figure 3. In the presence of O_2, glucose is converted to gluconic acid by the enzymes glucose oxidase and gluconolactonase (*37*). Gluconic acid (pK=3.6) is virtually fully dissociated at the pH values of relevance to this study. A third enzyme, catalase, can be added in the system to eliminate H_2O_2, which would otherwise contribute to the long term degradation of glucose oxidase (*38*).

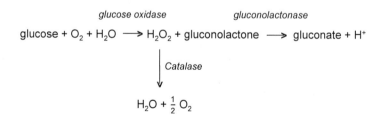

Figure 3. Schematic diagram of the enzymatic reaction.

Experimental section

Methods. Gel membranes consisting of (NIPA-*co*-methacrylic acid (MAA)) with ethylene glycol dimethacrylate (EGDMA) as crosslinker were synthesized in 1-4 dioxane at 60°C between two glass plates separated by 0.5 mm Teflon gaskets (*39*). 2,2'-Azobisisobutyronitrile was used as initiator. NIPA was purified before use following the procedure described by Gehrke et al. (*40*). The molar feed ratio of NIPA/MAA/EGDMA was 89/10/1 or 92/7/1 and dioxane mass was twice the total monomer mass. Membranes were conditioned in a series of methanol/water solvents and stored at 37°C for several days in a 50 mM NaCl solution containing 3 mM sodium azide (NaN_3).

Permeation experiments. Gels were mounted in a temperature controlled (37°C) side-by-side diffusion apparatus, with each cell having fluid volume of 75 mL (figure 4). For all the experiments the "donor" cell was pH-stated at 7.0 in order to simulate near-physiological conditions. First, the "receptor" cell's pH was varied in time using a pH-stat apparatus. Typically, the pH program in the receptor chamber started with pH sufficiently high (pH 5.3 or 5.4) that the gel was swollen. pH was then reduced in increments of 0.2 units every hour until a lowest value (pH 4.7 or 4.8) was reached, and pH was held there for two hours. pH was then increased every hour by 0.2 units until the end of the experimental run. Permeation of glucose across the gel membrane was measured by placing 1 μCi ^{14}C-glucose mixed with 1 mM cold glucose into the donor cell, also in the presence of 50 mM NaCl. In some cases 3 mM NaN_3 was also added. The receptor cell contained 50 mM NaCl, with or without NaN_3. Permeation was quantitated by ^{14}C counts in the receptor chamber.

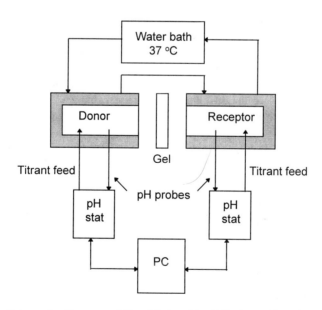

Figure 4. Schematic diagram of the side-by-side diffusion cell. Adapted from ref. (*39*) with permission.

In another experiment, the receptor chamber was kept for 20 hours at a pH sufficiently low that the membrane was in the collapsed state and then the pH was allowed to increase as a result of the passive diffusion of hydrogen ions to the donor cell. Permeation of glucose was quantitated as described above.

Measurement of pH oscillation. Glucose oxidase (21 IU/mL) with or without catalase (210 IU/mL) were added to receptor chamber. Cold glucose was added in the donor cell and the pH of the receptor cell was monitored with time. The specifications regarding the glucose concentrations in the donor cell are given in the figure captions. Bronopol (0.01% m/v) was used as preservative.

Results and discussion. Figure 5a shows the results of an experiment using a gel containing 10 mol% MAA, where pH was stepped down, in increments of 0.2 pH units, from pH 5.4 to pH 4.8, and then stepped back to pH 5.6. In this experiment NaN$_3$ was present. A steady permeation of glucose from donor to receptor is observed until pH passes from 5.0 to 4.8, at which point permeation becomes greatly attenuated. After holding at pH 4.8 and reversing the pH program, permeability remains very low until pH passes from 5.2 to 5.4, at which point permeability returns to its initial value. Figure 5b displays results of an identical experimental protocol, with the exception that the NaN$_3$ is absent. In this case shutdown of membrane permeability to glucose at pH 4.8 is incomplete, and appears to require a considerable delay. Full permeability to glucose is restored when pH sweeps upward past 5.4, as before. In figure 5c we show what occurs with NaN$_3$ absent, but pH now is lowered in 0.2 unit increments from 5.3 to 4.7, followed by incremental return to pH 5.5. When pH decreases from 4.9 to 4.7, glucose permeation becomes substantially attenuated. Full permeability is restored when pH sweeps past 5.3. Finally, in figure 5d where hydrogels containing 7 mol% MAA were used and in the absence of NaN$_3$, permeation decreases substantially at pH 4.8 as in figure 5a and the permeability returns to its initial value at pH 5.4. In all experiments, the differing pH values at which membrane shuts off and turns back on again are demonstrative of the pH-hysteresis we are seeking.

A similar experimental protocol was reported in reference (*39*). In that work NaN$_3$ was included in the system and was thought to be otherwise inert. A reproducible hysteretic behavior was also observed, with membrane shutoff occurring between pH 5.0 and 4.9, and restoration of permeability observed between pH 5.1 and 5.2. This hysteresis was robust with respect to the amount of time between transitions, suggesting that the hysteresis is a thermodynamic and not a kinetic phenomenon. The data in figure 5a are consistent with the results of reference (*39*). The more striking result is that the presence or absence of NaN$_3$ has a major impact on membrane behavior; compare figure 5a and 5b. This effect was not anticipated as NaN$_3$ was added as an antibacterial and was expected to be otherwise inert, especially at the low (3 mM) concentration. However, the presence of NaN$_3$ is required for membrane shutoff to occur at pH 4.8 since reduction of glucose permeability is much less and much slower at pH 4.8 in the absence of NaN$_3$. Without NaN$_3$, pH must be lowered to 4.7 for rapid shutoff to occur (figure 5c).

The gel membrane consists of a large amount of relatively hydrophobic comonomer (NIPA), and a smaller amount of titratable comonomer (MAA). Shutoff of the gel membrane when pH is lowered is attributed to formation of a dense skin layer proximal to the receptor cell, as a result of protonation and removal of charge from the MAA groups (3,15). Restoration of permeability occurs at higher pH due to deprotonation and recharging of these groups. If the amount of MAA in the hydrogel is lowered (figure 5d), membrane collapse and reswelling occur at a higher pH since the gel is more hydrophobic.

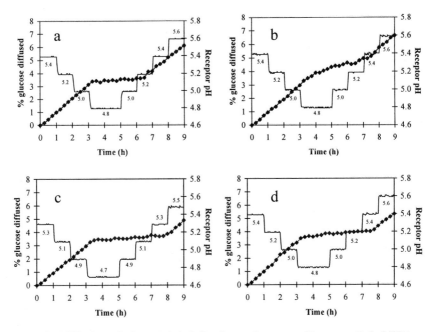

Figure 5. Diffusion of glucose (♦) following a down-up pH sweep. Poly(NIPA-co-MAA) gel containing 10 mol% (a,b,c) or 7 mol% MAA (d). Sodium azide 3 mM was added in experiment (a).

If shutoff is due to the protonation of MAA groups, then why does NaN_3 have such a strong effect? NaN_3 is the sodium form of hydrazoic acid, HN_3. The pKa for this acid is 4.7, and therefore it acts as a buffer in the pH region we are investigating. As a result larger additions of acid are needed to reduce the pH by a prescribed amount. The buffer then acts as a carrier for the excess protons and can deliver them to MAA groups, therefore accelerating gel collapse. Buffer-mediated acceleration of gel swelling and deswelling has been described by Chou et al. (41). These authors studied the influence of several buffers on the swelling of polybase and polyacid gels of 2-hydroxyethylmethacrylate, and found that the pKa of the buffer was a determining factor in the swelling kinetics. A second explanation is that the presence of buffer alters the pH-profile in the membrane. These two mechanisms need to be probed more quantitatively. Figure 5c shows that shutoff occurs at pH 4.7 even in the absence of azide. In this case, delivery of protons to the MAA groups in the gels is heightened simply by the enhanced availability of free protons at pH 4.7 compared to pH 4.8.

It has been established that gels which undergo discrete, first-order phase transitions do so in a manner that shows hysteresis. For instance, the swelling of poly(NIPA-*co*-SA) hydrogels and poly(NIPA-*co*-copper chlorophyllin-*co*-SA) exhibit hysteresis as a function of temperature (*4*) and pH (*42*), respectively. Here we provide indirect evidence that hysteresis in swelling may confer hysteresis in the permeability of a small molecule. While the precise mechanism behind the observed hysteresis is yet to be resolved, we speculate that the two primary forces (hydrophobicity and electrostatics) driving the transition lead to two free energy minima with a large kinetic barrier between them, and this barrier may be made higher by the shear modulus of the gel (*4,39,43*). A nonzero shear modulus can inhibit the nucleation of a swelling transition in a manner analogous to the inhibiting role that surface tension plays in the nucleation of liquid-vapor phase transitions, which leads to superheating and supersaturation phenomena. It should also be noted that the hysteretic behavior in a polyelectrolyte gel membrane may be affected by changes in the steepness of the pH gradient or the clamping conditions.

In order for a biochemical system not to reach equilibrium several criteria have to be met and the reader is referred to other reviews (*35,44*), but it can be shown that hysteresis substantially increases the potential of a system to oscillate (*27,34,45,46*). In this case the permeability of the membrane can switch back and forth from a high to a low state between the two critical pH values (*33*). The presence of buffering species alters the effective hysteresis band, and this effect needs to be taken into consideration when designing periodic, pulsatile drug delivery systems since in the body several physiological buffers are present.

Figure 6 shows the diffusion of glucose in the receptor cell as the pH gradually increase from 4.7 to 5.48 as a result of the diffusion of protons to the donor cell. During the first 6 hours the permeability of glucose is low since the membrane is in the collapsed state and increases sharply when the pH reaches 5.25, consistent with a first-order phase transition. The permeabilities of glucose (K_S) and protons (K_P) in the membrane were calculated using equations 4 and 5, respectively (*47*), and are reported in Table I for the 7 and 10 mol% MAA gels.

$$\frac{dQ}{dt} = \frac{K_S \, A Q_0}{V_D} \tag{4}$$

$$\frac{dpH_R}{dt} = \frac{K_P \, A}{2.303 \, V_R} \tag{5}$$

where

Q = cumulative mass of diffusant (glucose)
A = area for transport (3.14 cm^2)

Q_0 = initial amount of glucose added in the donor cell
V_D = volume of the donor cell (75 mL)
V_R = volume of the receptor cell (75 mL)
pH_R = pH in the receptor cell
t = time

Equation 5 is valid since the donor cell, at constant pH 7.0, can be considered a proton sink. We observe a 12-fold increase of glucose permeability when the membrane passes from the collapsed to the swollen state, while the permeability of protons remains relatively unchanged. Protons are apparently able to diffuse through pores in the collapsed membrane which are narrow enough to reduce glucose transport.

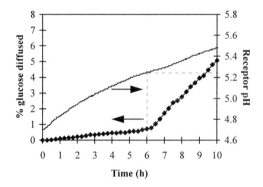

Figure 6. Diffusion of glucose (♦) following a gradual increase of pH.

Table I. Membrane permeability to glucose (S) and protons (P)

	Permeability (cm/s)	
K	7 mol% MAA	10 mol% MAA
$K_{S,H}$	7.8×10^{-5}	7.3×10^{-5}
$K_{S,L}$	6.6×10^{-6}	6.2×10^{-6}
$K_{P,H}$	8.5×10^{-4}	10.5×10^{-4}
$K_{P,L}$	11.7×10^{-4}	11.4×10^{-4}

In the proposed device, glucose diffuses into the receptor cell, where it is converted to protons which induce membrane shutdown. Then, a restorative process takes place because once the membrane has collapsed, the permeation of glucose is greatly attenuated while that of protons remains unmodified. Under these conditions, the pH in the receptor cell can rise until the membrane returns to its swollen state.

We have shown that it is possible to sharply diminish glucose permeability without altering the permeation of protons and provided evidence for pH-hysteresis. Thus all the elements needed to express a rhythmic behavior are present in this system.

To evaluate the oscillator, glucose oxidase and catalase were added in the receptor cell and a single dose of glucose (followed by another dose after 2 days to compensate for the enzymatic consumption of glucose) was placed in a closed donor cell. Again, the pH stat in the receptor cell was replaced by a pH meter. As shown in figure 7, the system exhibits multiple peaks in pH. However, attenuation of oscillations occurs with time. Indeed, it was found that the peaks, and especially the first one which far overshoots the limits of the hysteresis band, contain artifacts. At low pH we have found that impurities in the commercial preparation used reacts with protons over several hours (not shown) and this tends to accelerate the rate at which pH increases.

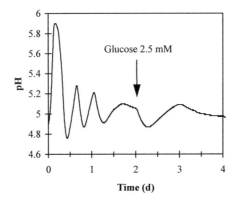

Figure 7. Variation of pH in the receptor chamber. 5 mM glucose was added in the donor cell followed by 2.5 mM after 2 days. The receptor cell contained 21 IU/mL crude glucose oxidase and 210 IU/mL catalase.

Accordingly, the catalase was removed and the glucose concentration in the donor chamber maintained constant by continuously circulating a stock solution of glucose. As shown in figure 8, the pH starts to oscillate in the hysteresis band but tends to reach a steady state after 4 days. The concentration of glucose used in this experiment was 1.5 mM. At higher (5 mM) or lower (0.8 mM) concentrations the peaks are more flattened (not shown). Using equation 3 and the calculated permeabilities of glucose and protons in the 10 mol% MAA gel (table 1), we find that the system is supposed to oscillate when the external glucose concentration is between 0.3 and 1.0 mM. Although these values are slightly different from the experimental data, this model provides a fair estimate of the glucose concentrations range that should yield oscillatory behavior.

Why is the system approaching steady state instead of oscillating indefinitely? First, observe in figure 8 that the time required for the pH to increase is long compared to figure 6, and the fully open state pH value (pH 5.25) is never reached. The slow increase in pH may be attributed to the residual diffusion of glucose in the shutoff state. Glucose can diffuse across the collapsed membrane and be converted to protons at a rate which significantly counters the depletion of protons from the receptor chamber. As argued elsewhere (Leroux and Siegel, submitted), this slow rate of pH variation may reveal the existence of intermediate swelling states of the membrane caused by phase separation heterogeneities in the structure of the gel (*48,49*), multiple phase transitions (*50*) or slight changes in swelling in the vicinity of the phase transition (*4*). Resolution of this issue requires further study. To optimize the device it appears necessary to better shut off the glucose permeation in the low permeability state by incorporating more hydrophobic comonomers in the gel network (*51*), and to decrease the oscillation period. The latter may be achieved by increasing membrane area or decreasing receptor chamber volume (*33*).

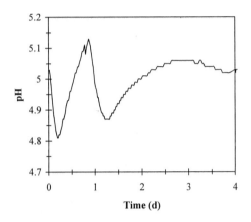

Figure 8. Variation of pH in the receptor chamber. The glucose concentration in the donor reservoir was maintained at 1.5 mM. The receptor cell contained 21 IU/mL glucose oxidase. Adapted from Leroux and Siegel (submitted).

As a final note, the present system should be distinguished from other enzyme/polymer systems which have been proposed to respond to *changes* in external physiologic concentration of a substrate, such as those which deliver pulses of insulin in response to increased glucose levels (*52-55*). While the system under study here shares with these other devices certain elements, such as a pH-sensitive hydrogel and the glucose oxidase enzyme system, its principle of operation (hysteresis-induced instability of negative feedback, driven by a *constant* level of external glucose) and its targeted set of drugs (e.g. hormones which require autonomous oscillatory delivery, which excludes insulin) set it apart from those previously considered constructs.

Conclusions

A scheme for a pulsatile drug delivery oscillator has been proposed. A simple model identifies the characteristics that are needed to achieve such behavior. In particular, the membrane permeability to substrate (glucose) needs to show hysteresis to product (H^+) concentration. We have also shown that a membrane with this property can be constructed. The response of these membranes depends to some degree on the

110

presence or absence of pH buffers, and this phenomenon impinges on the parameter specifications to be used in the theory. Preliminary evidence for oscillation is given. The device needs to be optimized but the presented data support the feasibility of the concept and open new perspectives for the periodic pulsatile delivery of drugs.

Acknowledgments

This work was funded by the NSF Grant CHF-96615511 and a postdoctoral fellowship to JCL from the Natural Sciences and Engineering Research Council of Canada. The authors acknowledge the advice of James Uchizono and John P. Baker.

References

1. Hoffman, A. S. In *Controlled drug delivery - Challenges and strategies*; K. Park, Ed. American Chemical Society: Washington, DC, 1997, pp 485-498.
2. Tanaka, T. *Physica* **1986**, *140A*, 261-268.
3. Hirotsu, S.; Hirokawa, Y.; Tanaka, T. *J. Chem. Phys.* **1987**, *87*, 1392-1395.
4. Sato Matsuo, E.; Tanaka, T. *J. Chem. Phys.* **1988**, *89*, 1695-1703.
5. Hoffman, A. S.; Afrassiabi, A.; Dong, L. C. *J. Controlled Release* **1986**, *4*, 213-222.
6. Brazel, C. S.; Peppas, N. A. *J. Controlled Release* **1996**, *39*, 57-64.
7. Yoshida, R.; Sakai, K.; Okano, T.; Sakurai, Y. *Adv. Drug Deliv. Rev.* **1993**, *11*, 85-108.
8. Yu, H.; Grainger, D. W. *J. Appl. Polym. Sci.* **1993**, *49*, 1553-1563.
9. Palassis, M.; Gehrke, S. H. *J. Controlled Release* **1992**, *18*, 1-12.
10. Tokuhiro, T.; Amiya, T.; Mamada, A.; Tanaka, T. *Macromolecules* **1991**, *24*, 2936-2943.
11. Okano, T.; Bae, Y. H.; Jacobs, H.; Kim, S. W. *J. Controlled Release* **1990**, *11*, 255-265.
12. Hirokawa, Y.; Tanaka, T. *J. Chem. Phys.* **1984**, *81*, 6379-6380.
13. Suzuki, A.; Tanaka, T. *Nature* **1990**, *346*, 345-347.
14. Park, T. G.; Hoffman, A. S. *Macromolecules* **1993**, *26*, 5045-5048.
15. Feil, H.; Bae, Y. H.; Feijen, J.; Kim, S. W. *Macromolecules* **1992**, *25*, 5528-5530.
16. Dong, L. C.; Hoffman, A. S. *J. Controlled Release* **1991**, *15*, 141-152.
17. Serres, A.; Baudys, M.; Kim, S. W. *Pharm. Res.* **1996**, *13*, 196-201.
18. Kim, Y. H.; Bae, Y. H.; Kim, S. W. *J. Controlled Release* **1994**, *28*, 143-152.
19. Parker, J. O.; Amies, M. H.; Hawkinson, R. W.; Heilman, J. M.; Hougham, A. J.; Vollmer, M. C.; Wilson, R. R. *Circulation* **1995**, *91*, 1368-1374.
20. Southworth, M. B.; Matsumoto, A. M.; Gross, K. M.; Soules, M. R.; Bremner, W. J. *J. Clin. Endocrinol. Metab.* **1991**, *72*, 1286-1289.
21. Santoro, N.; Filicori, M.; Crowley Jr, F. *Endocr. Rev.* **1986**, *7*, 11-23.
22. Wüthrich, P.; Ng, S. Y.; Fritzinger, B. K.; Roskos, K. V.; Heller, J. *J. Controlled Release* **1992**, *21*, 191-200.

23. Peppas, L. B. *J. Controlled Release* **1992**, *20*, 201-208.
24. Pekarek, K. J.; Jacob, J. S.; Mathiowitz, E. *Nature* **1994**, *367*, 258-260.
25. Giannos, S. A.; Dinh, S. M.; Berner, B. *J. Pharm. Sci.* **1995**, *84*, 539-543.
26. Giannos, S. A.; Dinh, S. M.; Berner, B. *Macromol. Rapid Commun.* **1995**, *16*, 527-531.
27. Gáspár, V.; Showalter, K. *J. Am. Chem. Soc.* **1987**, *109*, 4869-4876.
28. Rabai, G.; Hanazaki, I. *J. Phys. Chem.* **1996**, *100*, 10615-10619.
29. Yoshida, R.; Ichijo, H.; Hakuta, T.; Yamaguchi, T. *Macromol. Rapid Commun.* **1995**, *16*, 305-310.
30. Yoshida, R.; Yamaguchi, T.; Ichijo, H. *Mater. Sci. Eng.* **1996**, *C4*, 107-113.
31. Epstein, I. R. *J. Phys. Chem.* **1984**, *88*, 187-198.
32. Epstein, I. R.; Showalter, K. *J. Phys. Chem.* **1996**, *100*.
33. Siegel, R. A.; Zou, X.; Baker, J. P. *Proceed. Intern. Control. Rel. Bioact. Mater.* **1996**, *23*, 115-116.
34. Siegel, R. A. In *Controlled drug delivery - Challenges and strategies*; K. Park, Ed. American Chemical Society: Washington, DC, 1997, pp 501-527.
35. Murray, J. D. *Mathematical Biology* Springer-Verlag: Berlin, 1989.
36. Siegel, R. A.; Pitt, C. G. *J. Controlled Release* **1995**, *33*, 173-188.
37. Wilson, R.; Turner, A. P. F. *Biosensors Bioelectronics* **1992**, *7*, 165-185.
38. Tse, P. H. S.; Gough, D. A. *Biotechnol. Bioeng.* **1987**, *29*, 705-713.
39. Baker, J. P.; Siegel, R. A. *Macromol. Rapid Commun.* **1996**, *17*, 409-415.
40. Gehrke, S. H.; Palassis, M.; Akhtar, M. K. *Polymer Int.* **1992**, *29*, 29-36.
41. Chou, L. Y.; Blanch, H. W.; Prausnitz, J. M.; Siegel, R. A. *J. Appl. Polym. Sci.* **1992**, *45*, 1411-1423.
42. Suzuki, A.; Suzuki, H. *J. Chem. Phys.* **1995**, *103*, 4706-4710.
43. Sekimoto, K. *Phys. Rev. Lett.* **1993**, *70*, 4154-4157.
44. Edelstein-Keshet, L. *Mathematical Models in Biology* McGraw-Hill: New York, 1988.
45. Katchalsky, A.; Spangler, R. *Quart. Rev. Biophys.* **1968**, *2*, 127-175.
46. Otto, F. W.; Block, G. D. *Am. J. Physiol.* **1984**, *246*, R847-R851.
47. Flynn, G. L.; Yalkowsky, S. H.; Roseman, T. J. *J. Pharm. Sci.* **1974**, *63*, 479-510.
48. Sato Matsuo, E.; Orkisz, M.; Sun, S. T.; Li, Y.; Tanaka, T. *Macromolecules* **1994**, *27*, 6791-6796.
49. Li, Y.; Wang, G.; Hu, Z. *Macromolecules* **1995**, *28*, 4194-4197.
50. Annaka, M.; Tanaka, T. *Nature* **1992**, *355*, 430-432.
51. Yoshida, R.; Sakai, K.; Okano, T.; Sakurai, Y. *J. Biomater. Sci. Polymer Edn* **1994**, *6*, 585-589.
52. Kost, J.; Horbett, T.A.; Ratner, B.D.; Singh, M. *J. Biomed. Mater. Res.* **1984**, *19*, 1117-1133.
53. Ishihara, K.; Kobayashi, M.; Ishimaru, N.; Shinohara, I. *Polym. J.* **1984**, *16*, 625-631.
54. Fischel-Ghodsian, F.; Brown, L.; Mathiowitz, E.; Brandenburg, D.; Langer, R. *Proc. Natl. Acad. Sci. U.S.A.* **1988**, *85*, 2403-2406.
55. Makino, K.; Mack, E.J.; Okano, T.; Kim, S.W. *J. Controlled Release* **1990**, *12*, 235-239.

Chapter 9

Controlled Delivery Using Cyclodextrin Technology

Z. Helena Qi and Christopher T. Sikorski

Cerestar USA, Inc., 1100 Indianapolis Boulevard, Hammond, IN 46320

A great number of different processes are currently available for delivering active ingredients in a controlled fashion. New technologies are being developed constantly. Among them, cyclodextrin encapsulation has recently emerged as a novel technology complementary to the conventional approaches. Cyclodextrins are enzyme modified starch derivatives with a truncated-cone, or donut-shaped molecular structure. They are capable of forming inclusion complexes with a wide variety of organic and inorganic compounds on the basis of non-covalent interactions. Physical properties of the included molecules (guests) may thus be favorably modified. The basic concepts of cyclodextrin inclusion will be discussed, and the advantages of cyclodextrin applications will be illustrated.

Various controlled delivery technologies are used to provide some sort of protective barriers for an ingredient for controlling its delivery in intended use, enhancing its performance, increasing its safety or convenience, or prolonging its shelf-life (*1,2*). Controlled delivery products have many forms. They can have a discrete wall surrounding a core of active (*e.g.* a microcapsule). The active can be dispersed uniformly throughout the wall (*e.g.* a micromatrix). A matrix particle can have a protective wall built around it (*e.g.* microcapsule aggregate). A particle with an inert wall can contain the dissolved active as a molecular dispersion (*e.g.* microsphere). A cyclodextrin inclusion complex represents a different type of delivery vehicle that can offer unique advantages.

Cyclodextrins

Cyclodextrins are toroidally shaped, predominately rigid, cyclic oligosacaccharides consisting of D-glucopyranose units linked by α {1,4} glycosidic bonds.

Commercially available cyclodextrins consist of 6,7, or 8 D-glucose units and are referred to as α–,β–, and γ-cyclodextrins, respectively (Figure 1). Cyclodextrins are produced from starch by cyclodextrin-glycosyltransferase (CGTase), a process discovered in 1891 by A. Villiers (3). The structures of the cyclodextrins were elucidated by Schardinger between 1903 and 1911 (4,5,6). It is for this reason that cyclodextrins are sometimes referred to as Schardinger dextrins. In practice, individual cyclodextrins are produced either by using methods which direct the enzymatic conversion, or by using selective purification methods (7). The choice of enzyme and reaction conditions can also affect the distribution and yield of cyclodextrins formed. β-cyclodextrin is the predominant form produced by most enzymes, and it is also the most economical to produce at the present time. Significant commercial interest in cyclodextrins has arisen due to their ability to encapsulate various compounds since, in many cases, it is possible to beneficially modify the physical or chemical properties of the included guest molecules.

Physical and Chemical Properties

Molecular Structure. The relatively rigid nature of the cyclic structure constrains the primary and secondary hydroxyl groups to opposite rims of the cavity with the remaining C-H, C-C and C-O-C bonds lining the interior of the cavity. These structural features impart a hydrophilic and hydrophobic character to the exterior and interior of the cavity, respectively, and are responsible for the water solubility of cyclodextrins and their ability to encapsulate hydrophobic molecules within their cavities.

Solubility. α-, β–, and γ-Cyclodextrins have water solubilities of around 12.8%, 1.8%, and 25.6 % (w/w) , respectively, at ambient temperature (8). The water solubility of these cyclodextrins increase with increasing temperature. For example, β-cyclodextrin has a water solubility of approximately 10% (w/w) at 65 °C (9).

Hygroscopicity. α-, β–, and γ-Cyclodextrins have their equilibrium moisture content of around 10.2%, 14.5%, and 17.7% , respectively (10). The moisture levels approximately correspond to those typically found in starches. At these moisture levels, the cyclodextrins remain a pourable powder which is dry to the touch.

Chemical Stability. Cyclodextrins were originally termed by Villiers in 1891 as "cellulosine" due to their similarity to cellulose in being relatively resistant to hydrolysis (3). Cyclodextrins are more resistant to acid hydrolysis than starch, although strong acids such as sulfuric or hydrochloric acid can hydrolyze cyclodextrins (11,12). The rate of the hydrolysis is dependent upon temperature, the concentration, and type of acid. Cyclodextrins are typically not hydrolyzed by weak

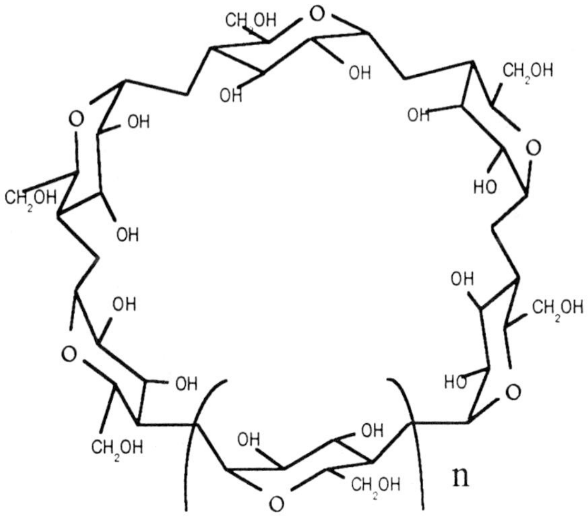

Figure 1.   Chemical structure of cyclodextrins (α-cyclodextrin where n = 1; β-cyclodextrin where n = 2; and γ-cyclodextrin where n = 3).

acids such as organic acids (*13*). Cyclodextrins are stable toward bases. No detectable hydrolysis of the cyclodextrin was observed in the presence of 0.35 N NaOH at 70 °C for 6 hours (*14*).

Thermal Stability. The thermal stability of cyclodextrins is far greater than that of starch. Two endothermic peaks are observed for α–, β–, or γ-cyclodextrins when they are analyzed by differential scanning calorimetry. The first peak occurs at roughly 100 °C owing to the evaporation of water. The second peak occurs at approximately 300 °C, corresponding to the simultaneous melting and decomposition of the cyclodextrin. In contrast, the thermograms obtained from starch typically exhibit additional peaks corresponding to changes in the secondary or tertiary structure of the starch molecules.

Enzymatic Stability and Metabolism. Cyclodextrins are not hydrolyzed by enzymes which require a reducing group, such as beta- and glucoamylases (*15*). However, they are susceptible to hydrolysis by various alpha-amylases (*16*). Cyclodextrins are also hydrolyzed by cyclodextrin glucosyltransferases, the enzyme which is used to produce cyclodextrins from starch (*17*). The metabolism of cyclodextrins is rather different. For instance, γ-cyclodextrin can be hydrolyzed as readily as starch (*18*), and only γ-cyclodextrin is rapidly degraded in the small intestines upon ingestion whereas α- and β-cyclodextrins are slowly degraded in the colon (*19*). Generally, α-cyclodextrin is the most stable and γ-cyclodextrins is the least stable toward enzymatic degradation among the three cyclodextrins. Chemical modification tends to decrease the biodegradability of cyclodextrin derivatives. It may be increasingly more difficult for a modified cyclodextrin to fit within the active site of enzymes as the degree of chemical substitution on the cyclodextrin increases.

Toxicology and Regulatory Status of β-Cyclodextrin. Numerous toxicological tests have been conducted on β-cyclodextrin (*20,21*). These tests include the Ames test, chromosomal aberration, teratology, Drosophila slrl, inhalation, and dermal and eye irritation. An oral LD_{50} value could not be established for β-cyclodextrin since no mortality was observed at the highest level of β-cyclodextrin incorporated into the diets. The caeca for the rats were enlarged at the 5% and 10% levels of β-cyclodextrin, but this is typical for slowly digestible carbohydrates. Beta-cyclodextrin at 0.1% and 0.3% levels was detected in the urine of rats fed at 5% and 10% levels of β-cyclodextrin, respectively, but no renal abnormalities were found.

Beta-cyclodextrin has been used in food applications outside the U.S. for years. For instance, β-cyclodextrin is regarded as a natural ingredient of food by the Ministry of Health and Welfare in Japan, and has been permitted without restriction in Japan since 1983. Various European countries have also approved, and are using β-cyclodextrin in food. The Joint FAO/WHO Expert Committee on Food Additives (JECFA) has recently recommended a maximum allowable daily consumption of up to 5 mg β-cyclodextrin/kg body weight (*20,21*). In the United States self-affirmation of β-cyclodextrin to be generally recognized as safe (GRAS) as a flavor protectant in human food has recently been completed (*22*). The petitioned maximum usage level is 2% β-cyclodextrin in ready-to-eat food.

Cyclodextrin Derivatives. Chemical modification of cyclodextrins can result in enhanced or expanded functionality of cyclodextrins. A significant amount of effort has focused on modifying the solubility of cyclodextrins, thus their complexed guest compounds. Typically, the relative solubility of the complex of a given guest, when made with different cyclodextrins, will parallel the solubility of the parent cyclodextrins. The solubility of cyclodextrin is usually dependent upon the hydrophobic or hydrophilic nature of the modified group, the degree of substitution, as well as the distribution of these groups on the cyclodextrin.

Cyclodextrins Derivatives with Increased Aqueous Solubility. Numerous derivatives have been reported which possess increased aqueous solubility relative to the respective unmodified cyclodextrins. Substitution on the hydroxyl groups disrupts the regular hydrogen-bonding network within a cyclodextrin molecule, and reduces the tendency of crystal formation, thus resulting in an increase in aqueous solubility (*23*). Most common cyclodextrin derivatives are produced through ether or ester linkages. For example, hydroxypropyl, and methyl ethers of β–cyclodextrin can be soluble in water up to 60% w/v or more (*24,25*). Acetylated β-cyclodextrins substituted with between 3-7 acetate groups per cyclodextrin are highly soluble as well (*26*). Ionic substitution groups such as carboxyalkyl (*27*), alkylamine (*27*), and sulfoalkyl (*28*) can provide additional hydration potential for high water solubility as well as electrostatic interactions with guest compounds bearing groups with opposite charges. In addition, crosslinked cyclodextrin polymers, with a MW range of 3,000 - 6,000 daltons, also exhibit higher aqueous solubility relative to the parent cyclodextrins (*29*). Derivatives possessing high aqueous solubilities are typically used in applications where it is desired to increase the solubility of a guest compound.

Derivatives with Decreased Aqueous Solubility. Alternatively, cyclodextrins can be chemically modified, typically by attaching hydrophobic groups extensively, to produce derivatives that are much less water soluble than their parents. For instance, fully acetylated cyclodextrins have aqueous solubilities which are typically below 1% by weight (*14*). High degree of polymerization, with epichlorohydrin or similar crosslinking reagents, can result in water insoluble beads of cyclodextrin polymers (*30*). A major application of these less soluble derivatives is as process aids where the cyclodextrin is used to remove certain components of interest from a mixture in the presence of water. The very low aqueous solubility of these derivatives means that little, if any, of the cyclodextrin will be left in the treated product, or that they may be used where a retarded or sustained release of the guest is desired. In addition, generally increased solubilities of these hydrophobic derivatives in non-aqueous media can further expand their application horizon over natural cyclodextrins.

Molecular Encapsulation Using Cyclodextrins. Encapsulation or complexation of a guest with cyclodextrins can be viewed as an equilibrium between bound and unbound forms, and the strength of the host-guest interaction can be expressed using the binding constant, K (*31,32*). It is also important to realize that the complexation process is reversible, particularly in aqueous solutions, and it is highly dynamic in the sense that there can be a continuous and rapid exchange between bound and unbound forms of the guest (Figure 2).

The nature of the guest and the morphology of the inclusion complex may be highly varied. For instance, the molecules which can be encapsulated by cyclodextrins are highly diverse in their physical properties, sometimes varying significantly in their degree of polarity or ionic character (*33,34*). The stoichiometry of the host:guest complex is typically one to one, although other ratios are known (*35*). The orientation of guests in the complex can also vary. For instance the encapsulated guest may be either axial (parallel) or equatorial (perpendicular) relative to the cyclodextrin cavity (*36,37,38*). In addition, molecules may only be partially included within the hydrophobic cavity owing to the hydrophilic nature of a portion of the molecule, or due to geometric incompatibility (*39*).

Driving Forces for the Cyclodextrin Complexation. The cyclodextrin encapsulation event can be governed by a number of energetic forces (*32*), such as (a) van der Waals and similar electrostatic interactions; (b) dipole-dipole and hydrogen bonding between guest and host; (c) the hydrophobic effect, which encompasses favorable solvent-solvent interactions which occur as a result of the association of hydrophobic molecules due to the "squeezing-out" effect of water and (d) the expulsion of high energy water molecules from the hydrophobic cyclodextrin cavity. The relative importance of these so-called 'weak forces' is dependent upon the guest, the cyclodextrin, and the solvent, and in most cases more than one force will contribute collectively to the complexation process.

Factors Affecting Inclusion Complexation. Since the cavity size of a cyclodextrin varies with the number of glucose units in the ring, the geometric compatibility of a cyclodextrin with a guest compound is of primary importance (*40*). This single factor will not only determine whether an inclusion complex can be formed, but also the stability of the complex, if it does form. In addition, the binding strength of a given complex can also be affected by a variety of factors, which permit a way to control the release of the encapsulated guest. For instance, the presence of competitive guests for the cyclodextrin cavity can shift the equilibrium of the guest more towards the free state (*41*). In addition, heating in aqueous solution can be used to destabilize the complex (*42*). The pH level of the solution may influence the cyclodextrin-guest binding by changing the ionic state of the guest or the cyclodextrin. For example, deprotonation of the secondary hydroxyl groups of the cyclodextrins at high pH conditions can also destabilize complexes.

$$CD + Guest \xrightarrow{\quad K \quad} CD\text{-}Guest$$

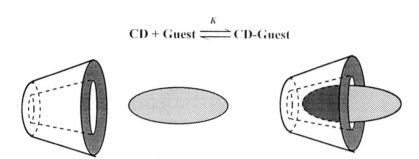

Figure 2. Schematic illustration of the dynamic complexation equilibrium between a cyclodextrin (CD) and a guest molecule (Guest).

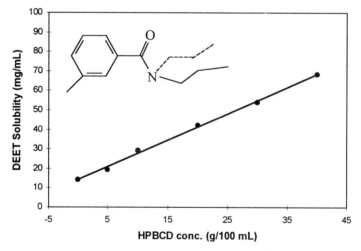

Figure 3. Solubilization of DEET (N,N-diethyl-3-methylbenzamide) using hydroxypropyl-β-cyclodextrin (HPBCD).

Effects of Cyclodextrin Encapsulation

Alteration of Chemical Properties. Although cyclodextrins have been used as catalysts, even enzyme models in academic studies (*43*), where significant acceleration of reaction rates are achieved, in practical applications it is the stabilization effect of cyclodextrins that is most valued and utilized.

Many active ingredients are chemically labile which are sensitive to light, heat, or oxidation. Inclusion of a guest in the cavity of cyclodextrin can shield the guest molecule from various destabilizing factors. For example, vitamin A degrades rapidly when it is exposed to light. In a stability study on vitamin A palmitate (*14*), the photodegradation half-life of the vitamin was measured to be about 1.6 hours at room temperature. Upon forming a complex with ß-cyclodextrin, the half-life increased significantly to around 45 hours. Benzaldehyde is a component of many flavors and it has also been found to have antitumor activity (*44*). However, it is an oily liquid with sparing solubility in water. Its instability to air and light presents considerable problems in its food and pharmaceutical applications. In order to overcome these difficulties the complexes of benzaldehyde with all three cyclodextrins have been prepared. In one study (*45*), the autooxidation rates of free benzaldehyde and its inclusion complex with α-cyclodextrin in neutral solution were determined at 70°C. The half-life of the complex was 3.6 times that of the free guest. In another experiment, the oxidation of benzaldehyde and its complex with β-cyclodextrin were measured in pure oxygen atmosphere using a Warburg apparatus (*46*). During a period of time when observations were made, while the complexed benzaldehyde remained virtually intact, the free guest was rapidly oxidized to yield the inactive benzoic acid.

Modification of Physical Properties. The association of a guest molecule with a cyclodextrin cavity can result in marked changes of its physical properties. Although numerous studies have been published in this regard, examples of the cyclodextrin inclusion-induced modifications will be limited to those that have existing or potential commercial applications.

Aqueous Solubility. The water solubility of guests can be altered by complexation with cyclodextrins, particularly their derivatives. Due to the interaction of the hydroxyl groups with water, cyclodextrins themselves have sufficient hydrophilicity to be soluble in water. When a highly water insoluble compound is included inside the cavity of cyclodextrin, the hydrophobic moiety is surrounded by the cyclodextrin. Hydrophilic derivatives are generally much more effective solubilizers than their parent cyclodextrins. DEET (N,N-diethyl-3-methylbenzamide) is a widely used insect repellent which is only sparingly soluble in water. However, the solubility of DEET increased linearly with the concentration of hydroxypropyl-β-cyclodextrin (Figure 3).

On the other hand, decrease in water solubility of some hydrophilic molecules upon complexation with β-cyclodextrin or some hydrophobic cyclodextrin derivatives has been utilized as a means of controlled or sustained release.

Enhancement of Physical Stability. When a volatile guest compound is included inside the cavity of a cyclodextrin, the individual guest molecules can be 'compartmentalized' enough so that the melting or boiling profiles change greatly. For instance, many liquid flavorings and fragrances have been made and marketed in their solid forms as cyclodextrin complexes. Even more dramatically, a gaseous substance may be converted to a powder upon inclusion by cyclodextrins, as in the case of CO_2-αCD complex (*47*). This effect of cyclodextrin complexation has practical applications. Menthol, a volatile solid at room temperature, for example, has been stabilized by cyclodextrins, as evidenced by shifting of both its melting and boiling temperatures (*48*). Similar results have been obtained for methyl salicylic acid.

Modification of Sensory Characteristics. Sensory characteristics, such as taste, smell, and local irritant reactions of certain guest compounds can also be altered upon inclusion. Thus, taste of food and orally administered drugs may be improved by masking off-flavors and unpleasant odors through cyclodextrin complexation. Dextromethorphan, an antitussive drug formulated in many over-the-counter cold medicines, has an astringent bitter taste. Although being a bulky molecule (Figure 4), Dextromethorphan hydrobromide forms an inclusion complex with β-cyclodextrin with 1:1 stoichiometry, and its stability constants at pH 7.4 and pH 4.2 are, respectively, 8000 ± 800 M^{-1} and 5750 ± 500 M^{-1}(*49*). Previous findings indicated that the bitterness of both Dextromethorphan base (Dm) and Dextromethorphan hydrobromide (Dm•Br) could be reduced when they were in the form of β-cyclodextrin complexes (*50)*. In our study significant differences (P < 0.05) were identified among free Dm• Br, Dm• Br, its1:1 and 1:2 molar equivalents of β-cyclodextrin by 25 panelists (Table I). Moreover, in aqueous test solutions direct addition of β-cyclodextrin prior to mixing was equally effective as prior formation of a Dm• Br /β-cyclodextrin complex.

The non-steroidal anti-inflammatory drugs (NSAIDs) are one of the most commonly used group of oral drugs. However, most NSAIDs cause some form of gastro-intestinal irritation. One of several approaches to overcome the problem is the use of cyclodextrin technology. Complexation of NSAIDs with cyclodextrins often results in more rapid absorption of the drug after oral administration, in addition to the stabilizing effect, thus reducing the potential for gastric lesions. For example, significantly increased gastric tolerance of the piroxicam-β-cyclodextrin complex (*51*) has led to its commercialization in many countries including France, Germany, Belgium, Brazil, Korea, Pakistan and Italy (*52*).

Use of Cyclodextrins as Process Aids. Cyclodextrins have also been used as process aids to encapsulate and remove specific components from mixtures. The encapsulated material can be isolated, and separated, thereby allowing recovery of the removed components and the cyclodextrin. The components can then be purified further, if necessary, and the recovered cyclodextrin can be recycled.

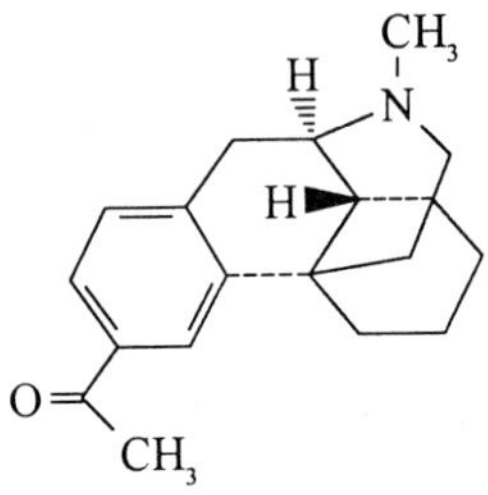

Figure 4.   Chemical structure of dextromethorphan.

Table I. **Sensory Evaluation Results for Dextromephorphan Hydrobromide (Dm) Solutions in the Absence and Presence of β-cyclodextrin (βCD).**

Panel Number	Solution A Dm (2 mg/ml)	Solution B βCD/Dm (1:1)	Solution C βCD/Dm (2:1)
1	3	2	1
2	3	2	1
3	3	2	1
4	3	2	1
5	3	1	1
6	2	1	1
7	3	2	1
8	3	3	0
9	3	2	1
10	3	2	1
11	3	2	1
12	3	2	1
13	3	2	0
14	3	2	1
15	3	2	0
16	3	2	1
17	3	3	0
18	3	2	2
19	3	1	1
20	3	2.5	2.5
21	3	2	1
22	3	2.7	0
23	3	2	3
24	3	1	1.5
25	3	2	2
X (average)	3.0	2.0	1.0

Notes:

1. The intensities of bitterness were ranked according to the scale of: 0 = not bitter, or very slightly bitterer; 1 = slightly bitter; 2 = bitter; 3 = very bitter.

2. Each solution contained 2 mg/ml of the free drug. Additional β-cyclodextrin (2 molar equivalent) was present in Solution C, whereas β-cyclodextrin (1 molar equivalent) and an inert-tasting dextrin (at the weight of 1 molar equivalent of β-cyclodextrin) were present in Solution B.

Beta-cyclodextrin has been used to remove cholesterol from egg yolk. Egg yolk or whole eggs were treated with β-cyclodextrin according to a method developed by Cerestar and Michigan State (53). Approximately 80% of the cholesterol was removed from the egg product using 6% β-cyclodextrin (w/w) relative to the egg yolk at a mixing time of less than 10 minutes at 55 °C. The pH of the yolk mixture is adjusted with base in order to solubilize protein materials during the process. The cyclodextrin-cholesterol complex is removed as an insoluble paste using centrifugation. The pH of the egg product is then re-adjusted with acid to the desired pH range. Residual sodium or potassium level in the product, arising from pH adjustments, typically amounts to less than 0.25% by weight. This level of residual salt does not adversely effect the sensory properties of the egg product. Residual β-cyclodextrin is removed from the product with a water-insoluble immobilized amylase, thereby providing a product which is free of either cyclodextrin or enzyme (42). Egg yolk recovery for the entire process is about 85%, and the recovered product is indistinguishable in functionality, flavor, and taste compared to untreated egg yolk. The cyclodextrin-cholesterol complex can also be separated by heating which allows recovery of about 85% of the β-cyclodextrin. The cholesterol removing efficiency of the recovered cyclodextrin is virtually the same as the virgin β-cyclodextrin.

Cyclodextrins for Controlled Release

Controlled release of the encapsulated guests is of great importance in delivering active ingredients at the site of action. In general there are three types of controlled-release modes, accelerated (or enhanced), retarded (or sustained) and triggered (or timed Selective use of different types of cyclodextrin derivatives can often facilitate in achieving desirable control over the guest release (54,55).

Accelerated/Enhanced Release. Examples of accelerated release of drugs in different dosage forms are abundant in the literature. Some of them are listed in Table II. Several effects of the cyclodextrin inclusion can contribute to the overall enhanced bioavailability of the drugs. These effects include increases in the solubility, dissolution rate, as well as release rate. Best acceleration result is achieved by using cyclodextrin derivatives with increased aqueous solubility, such as hydroxypropyl cyclodextrins and methyl cyclodextrins, for lipophilic compounds with relatively low complex stability. For example, when equivalent doses of α-tocopheryl nicotinate were orally administered to fasting dogs, the drug in the form of dimethyl-β-cyclodextrin complex gave a maximal plasma level about 17 times higher than that of the free drug. The area under the plasma concentration/time curve (AUC) of the complex was about 70 times as great as that of the free drug (56).

Retarded/Sustained Release. Table III contains a few examples and their literature references related to this type of release. As a cooling agent, 3-1-menthoxypropane-1,2,-diol, is used in chewing gums. The desired 'coolness' sensation peaks at the initial chewing, and diminishes during the later stages of the chew. In addition, at its required level the cooling agent causes an undesirable softening of the gum base, which is both inconvenient and expensive to overcome by replacement with a modified gum base. By the use of 3-1-menthoxypropane-1,2,-diol in the form of a cyclodextrin complex, however, both problems were solved (*63*). The chewing gum containing the cyclodextrin complex provided a sustained release of the cooling flavor. At the same time, the gum base softening effect caused by the cooling agent was minimized.

Table II.   Examples of Accelerated Release of Guests by Cyclodextrins

Guest Compound	Cyclodextrin	Application	Ref.
prostaglandin E1	methyl-β	pharmaceutical (oral; tablets)	57
theophylline	β	pharmaceutical (oral; tablets)	58
isosorbide dinitrate	hydroxypropyl-β	pharmaceutical	59
nifedipine	hydroxypropyl-β	pharmaceutical	60
menaquinone	methyl-β	pharmaceutical (oral)	61
carmofur	methyl-β	pharmaceutical (oral)	62

Cyclodextrin derivatives with decreased aqueous solubility are advantageous in retarding the release of their included hydrophilic guests into an aqueous medium. Matsubara has reported the use of peracetylated cyclodextrins to encapsulate the buserelin acetate, an agonist of luteinizing hormone-releasing hormone (*64*). The encapsulated drug can be injected in an oil based suspension, and is released in a sustained fashion.   A single subcutaneous injection of the oily suspension of the drug containing acetylated β- and γ-cyclodextrins led to retardation of plasma levels of buserelin acetate, resulting in 25- and 39-fold longer mean residence times, respectively, than that of the drug alone.

Triggered/Timed Release. In certain applications, it is ideal that the active ingredient can be released in response to a 'trigger' at a desired time or location to perform its intended function. While a higher degree of control is obviously required to achieve this objective, encouraging results have been obtained by use of cyclodextrins. A good example is the one part heat curable organopolysiloxane system developed by Lewis *et. al.* (*82*). The cyclooctadiene platinum catalyst in the mixture was greatly stabilized in the form of a β-cyclodextrin complex, which had a shelf stability of more than 7 months at 25°C compared with that of 12 hours shown by the uncomplexed catalyst. When the hydrosilation polymer was heated to 150 °C, however, the heat triggered a quick release of the catalyst to cure the silicone.

Another type of 'release trigger' is the pH of the medium, which may influence the binding of a guest compound with a suitable ionic cyclodextrin derivative.   For instance, CMEBCD is produced through the introduction of carboxymethyl groups to the remaining hydroxyls of ethylated β-cyclodextrin, which confers on the included guest a pH-dependent solubility. CMEBCD was used as a model carrier for drugs that are (a) unstable in the stomach, (b) absorbed mainly from the intestinal tract, and (c) irritable to the gastric mucus (*54*). Diltiazem hydrochloride, a potent calcium antagonist, is absorbed primarily from the jejuno-ileum, and has a short biological half-life. It was found that the release rate of diltiazem from the starch tablet was very fast and independent of the pH of the medium. On the other hand, the release rate from the tablet containing the CMEBCD complex was slow at low pH regions, but accelerated with an increase in the pH of the medium, depending on the ionization of the carboxymethyl groups of the host cyclodextrin. After oral administration of the diltiazem tablet containing the CMEBCD complex to dogs, a nearly twofold increase in bioavailability was observed compared with the free drug. Additional examples and their reference can be found in Table IV.

Table III.   Examples of Sustained Release of Guests by Cyclodextrins

Guest Compound	Cyclodextrin	Application	Ref.
α-halocinnamaldehydes	β	microbicide	65
cetyl-pyridinium chloride	β	antiseptic or disinfectant	66
maleic acid, fumaric acid	β	fungicide	67
iodine	β	wound treatment	68
fragrances (various)	β	cosmetics and coatings	69
diltiazem	hydroxyethyl-β	pharmaceutical (oral)	70
prostaglandin E1	hydroxyethyl-β	pharmaceutical (oral; tablets)	57
theophylline	hydroxyethyl-β	pharmaceutical (oral; tablets)	58
estradiol	β	pharmaceutical (transdermal)	71
griseofulvin	crosslinked β polymer	pharmaceutical	72
morphine hydrochloride	crosslinked β polymer	suppository	73
hydrocortisone	hydroxypropyl-β	drug microcapsules	74
diazepam	hydroxypropyl-β	pharmaceutical (erodible matrixes)	75
1,7-dioxaspiro[5,5] undecane	β	pesticide	76
α-pinene	β	pesticide	76
5-fluorouracil	α	suppository	77
nifedipine	carboxymethyl-ethyl-β	pharmaceutical	60
carboplatin	hydroxypropyl-α	pharmaceutical	78
molsidomine	per-acylated-β	pharmaceutical	79
salbutamol hydrosulfate	per-acylated-β	pharmaceutical	80
pilocarpine	β	pharmaceutical	81

Table IV.   Examples of Triggered Release of Guests by Cyclodextrins

Guest Compound	Cyclodextrin	Application	Ref.
perfume	β	household (fabric treatment)	83
theophylline	carboxymethylethyl-β	pharmaceutical (oral; tablets)	58
aromatic perfumes	carboxymethylethyl-β	aromatherapy	84
molsidomine	carboxymethylethyl-β	pharmaceutical	85
5-aminosalicylic acid	β	pharmaceutical	86
isothiocyanate	β	antimicrobial food packaging	87

Conclusion

Over 100 years after their discovery, cyclodextrins have become a commercial reality. Great potential for these cyclic molecules in commercial applications has been gradually unveiled since the late 1970s. Their ability to form molecular inclusion complexes, therefore altering the properties of their included materials, allow them to find usage in such widespread areas that include most industries involving organic chemistry. In the arena of controlled delivery, the cyclodextrin technology has demonstrated its unique advantages, with preliminary results, as a novel means complementary to the conventional encapsulation methods. As their production improves, more cyclodextrin research advances are made, new applications discovered, and the regulatory hurdles crossed, the next decade will surely witness a continued surge in the commercialization of products utilizing cyclodextrins.

References

1.   DiCicco, R. L. *Proceed. Intern. Symp. Control. Rel. Bioact. Mater.* **1996**, *23*, 127.
2.   Dziezak, J. D. *Food Technology*; April-**1988**, 136-157.
3.   Villiers, A. *Compt. Rend. Acad. Sci. Paris* **1891**, *112*, 536.
4.   Schardinger, F. *Unters. Nahrungs-, Genusmitelel, Gebrauchsgegenstande* **1903**, *16*, 865.
5.   Schardinger, F. *Wien, Klin. Wochenschr.* **1904**, *17*, 204.
6.   Schardinger, F. *Zentr. Bakteriol. Parasitenk. II* **1911**, *29*, 188.
7.   Szejtli, J. *Cyclodextrin Technology* ; Ed. Davies, J. E. D., Ed; Kluwer Academic Publishers: Dordrecht, **1988**, pp.34-39.
8.   Jozwiakkowski, M.J.; Connors, K. A. *Carbohydr. Res.* **1985**, *143*, 51.
9.   Wiedenhof, N.; Lammers, J. N. N. J. *Carbohydr.* Res. **1968**, *7*, 1.
10.   Szejtli, J. *Ref. (7)*, pp.12.
11.   Swanson, M.A.; Cori, C. F. *J. Biol. Chem.* **1948**, *172*, 797-804.
12.   Myrback, B. K.; Jarnestrom, T. *Kami* **1949**; *1*; 129-143.

128

13. Schonberger, B.P.; Jansen, A. C. A.; Jannssen, L. H. M. *Proceedings of the Fourth International Symposium on Cyclodextrins*; Huber, O.; Szejtli, J., Eds.; **1988**, pp.61.
14. Unpublished results (Cerestar USA, Inc.).
15. Szejtli, J. *Ref. (7)*, pp.24.
16. Szejtli; J. *Cyclodextrins and Their Inclusion Complexes*; Academiai Kiado: Budapest, **1982**, pp.42-48.
17. French, D.; Levine, M. L.; Norberg, E.; Norain, P.; Pazur, J. H.; Wild, G. M. *J. Am. Chem. Soc.* **1954**, *76*, 2387.
18. Jodal, I.; Kandra, L.; Harangi, J.; Nanasi, P.; Szejtli, J. *Starch* **1984**; *36*, 140.
19. Gergely, V.; Sebastion, G.; Geyen, G.; Virag, S. *Proceedings of the First International Symposium on Cyclodextrins*; Szejtli, J., Ed.; Academiai Kiado: Budapest, Hungary, **1981**, pp.109-133.
20. *Toxicological Evaluation of Certain Food Additives and Contaminants* **1993**, WHO Food Additive Series no. 32, WHO, Geneva.
21. *Toxicological Evaluation of Certain Food Additives and Contaminants* **1996**, WHO Food Additive Series no. 35, WHO, Geneva.
22. *Food Chemical News*, Nov.3, **1997**.
23. Szejtli, J. *Ref. (7)*, pp.4.
24. Pitha, J. *US Pat.* **1988**, 4,727,064.
25. Tsuchiyama, Y.; Sato, M.; Yagi, Y.; Ishikura, T. *US Pat.* **1988**, 4,746,734.
26. Liu, F.-Y.; Lildsig, D. O.; Mitra, A. K. *Drug Dev. Ind. Pharm.* **1992**, *18(15)*, 1599-1612.
27. Parmerter, S. M.; Allen, E. E.; Hull, G. A. *US Pat.* **1969**, 3,426,011
28. Stella, V.; Rajewski, R. *US Pat.* **1992**, 5,134,127.
29. Fenyvesi, E.; Szilasi, M.; Zsadon, B.; Szejtli, J.; Tudos, F. *Proceedings of the First International Symposium on Cyclodextrins*; Szejtli, J.,Ed.; Academiai Kiado: Budapest, Hungary, **1981**, pp.345.
30. Wiedenhof, N. *Die Starke* **1969**, *6*, 163-166.
31. Connors, K. A. *Binding Constants* John Wiley & Sons: New York, **1987**, pp.21-30.
32. Connors, K.A. *Chem. Rev.* **1997**, *97*, 1325-1357.
33. Cramer, F.; Henglein, F. M. *Chem. Ber.* **1957**, *90*, 2561.
34. Matsui, Y.; Nishioka, T.; Fujita, T. *Top. Curr. Chem.* **1985**, *128*, 61.
35. Bender, M.; Komiyama, M. *Cyclodextrin Chemistry* Springer-Verlag: New York, **1978**, pp.102-103.
36. Kobyashi, N. *J. Chem. Soc., Chem. Commun.* **1989**, 1126.
37. Harata, K.; Uedaira, H. *Bull. Chem. Soc. Jpn.* **1975**, *48*, 375.
38. Shimizu, H.; Kaito, A.; Hatano, M. *J. Chem. Soc.* **1982**, *104*, 7059.
39. Divakar, D.; M. Mahesaran, M. M. *J. Inclusion Phenom. Mol. Recognit. Chem.* **1997**, *27*, 113-126.
40. Szejtli, J. *Ref. (7)*, pp. 80-81.
41. Szejtli, J. *Ref. (7)*, pp. 153-154.
42. Shieh, W.; Hedges, A. *U.S. Pat.* **1996**, 5,565.226.
43. Szejtli, J. *Ref. (7)*, pp. 117-121.

44. Sakaguchi, R.; Hayase, E. *Agric. Biol. Chem.* **1979**, *43*, 1775.
45. Chang, C-J.; Choi, H. S.; Wei, Y. C.; Mak, V.; Knevel, A. M.; Madden, K. M.; Carlson, G. P.; Grant, D. M.; Diaz, L.; Morin, F. G. *Biotechnology of Amylodextrin Oligosaccharides, ACS Symposium Series 458*; Friedman, R. B, Ed.; American Chemical Society, Washington, D. C., **1991**
46. Szejtli, J.; Szente, L.; Bánky-Elöd, E. *Acta Chim. Acad. Sci. Hung.* **1979**, *101*,
47. House Food Ind. K.K., *Japan Kokai*, **1976**, 51,148,052.
48. Qi, Z. H.; Hedges, A. H. *Use of Cyclodextrins for Flavors*, in *Flavor Technology, ACS Symposium Series 610*, Ho, C.-T.; Tan, C.-T.; Tong, C.-H., Eds.; American Chemical Society, Washington, D. C., **1995**
49. Thuaud, N.; Gosselet, N.-M.; Sebille, B., *J. Inclusion Phenom. Mol. Recognit. Chem.* **1996**, *25*, 267-281.
50. Szejtli, J. *Ref (7)* pp.225.
51. Santucci, L.; Fiorucci, S.; Patoia, L.; Farroni, F.; Sicilia, A.; Chiucchiu, S.; Bufalino, L.; Morelli, A. *Drug Invest.* **1990**, *2 (Suppl. 4)*, 484-490.
52. *Cyclodextrin News*; Szejtly, J., Ed.; **1997**, *11. N°5*, 92.
53. Awad, A; Hedges, A.; Shieh, W.; Sikorski, C.; Smith; D.M. *WO 9629893 AOPAB:961111*, **1996**.
54. Uekama, K.; Hirayama, F.; Irie, T. *New Trends in Cyclodextrins and Derivatives*; Duchêne, D., Ed.; Edition de Senté, Paris, **1991**, 409-446.
55. Qi, Z. H. *Polymer News*, to be published in 1998.
56. Uekama, K.; Horiuchi, Y.; Kikuchi, M.; Hirayama F., Ijitsu, T.; Ueno, M. *J. Inclusion Phenom. Mol. Recognit. Chem.* **1988**, *6*, 167-174.
57. Uekama, K.; Kurihara, M.; Hirayama, F. *Nippon Kagaku Kaishi* **1990**, *10*, 1195-1199.
58. Horiuchi, Y.; Abe, k.; Hirayama, F.; Uekama, K. *J. Contr. Rel.* **1991**, *15(20)*, 177-183.
59. Seo, H.; Oh, K.; Hirayama, F.; Uekama, K. *The Sixth International Symposium on Cyclodextrins*; Hedges, A. R., Ed.; Edition de Senté :Paris, France, **1992**, pp.543-546.
60. Wang, Z.; Horikawa, T.; Hirayama, F.; Uekama, K. *Proceedings of the Seventh International Symposium on Cyclodextrins* **1994**, pp.439-444.
61. Horiuchi, Y.; Kikuchi, M.; Hirayama, F.; Uekama, K.; Ueno, M.; Ijitsu, T. *Yakugaku Zasshi* **1988**, 1093-1100.
62. Kikuchi, M.; Uekama, K. *Xenobiotic Metabolism and Disposition* **1988**, *3*, 267-273.
63. Patel, M. H.; Hvizdos, S. A. *U.S. Pat.* **1992**, 5,165,943.
64. Matsubara, K.; Irie, T.; Uekama, K. *J. Contr. Rel.* **1994**, *31*, 173-80.
65. Kawachi, T. *Jpn. Kokai Tokkyo Koho* **1985**, JP 60188302 A2.
66. Friedman, R. B. *US Patent* **1988**, 4,774,329.
67. Furukawa, M.; Hara, K. *Japanese Patent* **1988**, 88,83,003.
68. Szejtli, J.; Fenyvesi, E.; Sarkozi, P.; Felmeray, I.; Zsoldos, A. *German Patent*, **1988**, 3,819,488.
69. Chikahisha, N.; Cho, S. *Japanese Patent* **1989**, 89,40,567.

70. Uekama, K.; Hirayama, F.; Irie, T. *Proceedings of the Fifth International Symposium on Cyclodextrins*; Duchene, D., Ed.; Edition de Senté :Paris, France, **1990**.pp.457-482.

71. Hansen, J.; Mollgaard, B. *PCT Int. Appl.* **1991**, WO 9109592 A1.

72. Carli, F.; Colombo, I.; Ragaglia, L. *Eur. Pat. Appl.* **1991**, EP 446753 A1.

73. Tobino, Y.; Torii, H.; Ikeda, K.; Nakamura, K.; Arima, H.; Irie, T.; Uekama, K. *Kyushu Yakugakkai Kaiho* **1991**, *45*, 15-20.

74. Loftsson, T.; Kristmundsdottir, T.; Ingvarsdottir, K.; Olafsdottir, B. J.; Baldvinsdottir, J. *J. Microencap.* **1992**, *9(30)*, 375-382.

75. Conte, U.; Giunchedi, P.; Maggi, L.; La Manna, A. *S.T.P. Pharma Sci.* **1993**, *3(3)*, 24209.

76. Mazomenos, B.; Hadjoudis, E.; Yannacopoulou, C.; Botsi, A.; Tsoucaris, G. *Eur. Pat. Appl.* **1993**, EP 572743 A1.

77. Koyama, Y.; Kasama, T.; Noguchi, Y. *Jpn. Kokai Tokkyo Koho* **1994**, JP 06040890 A2.

78. Loftsson, T.; Pitha, J.; Kristmundsdottirm T.; Utsuki, T.; Olivi, A.; Brem, H. *Proceedings of the Seventh International Symposium on Cyclodextrins* **1994**, pp.423-426.

79. Uuekama, K.; Horikawa, T.; Yamanaka, M.; Hirayama, F. *J. Pharm. Pharmacol.* **1994**, *46(6)*, 714-717.

80. Hirayama, F.; Uekama, K. *Pharm. Tech. Jpn.* **1995**, *11(1)*, 19-24.

81. Davies, N. M.; Barry, A. R.; Tucker, I. G. *Proc. Int. Symp. Controlled Release Bioact. Mater.* **1996**, 719-720.

82. Lewis, L. N.; Sumpter, C. A. *U.S. Pat.* **1991**, 5,025,073.

83. Trinh, Toan; Gardlik, J. M.; Banks, T. J.; Benvegnu, F. *Eur. Pat. Appl.* **1990**, EP 392607 A1.

84. Fuwa, T.; Uekama, K. *US Patent* **1993**, 5,238,915.

85. Horikawa, T.; Hirayama, F.; Uekama, K. *J. Pharm. Pharmacol.* **1995**, *47(20)*, 124-127.

86. Weckenmann, H. P.; Siehka, V.; Bauer, K. H. *Proc. of the 14th Pharm. Technol. Conf.* **1995**, 390.

87. Tachika, S. *Jpn. Kokai Tokkyo Koho* **1995**, JP 07265027 A2.

Chapter 10

Inclusion Compounds as a Means to Fabricate Controlled Release Materials

L. Huang and A. E. Tonelli

Fiber and Polymer Science Program, College of Textiles, North Carolina State University, Campus Box 8301, Raleigh, NC 27696–8301

Certain molecular hosts, such as urea, thiourea, perhydrotriphenylene, and cyclodextrins, can form crystalline inclusion compounds (ICs) during their cocrystallization with appropriate guest molecules. The IC host molecules crystallize into a three-dimensional lattice which surrounds and isolates the included guest molecules into well-defined cavities. These IC crystals may be thought of as host molecule crystalline containers whose contents are the included guest molecules. Until the IC crystals are disrupted by melting or dissolution, the included guest molecules are kept isolated from the environment. Both small-molecule and polymer guests may be included in ICs. When embedded in a carrier polymer phase and subsequently treated with a solvent for the IC host, the included guest molecules are released and coalesced into the carrier polymer phase producing a guest-carrier polymer molecular composite. In the present report we describe the fabrication of several such molecular composites using ICs containing either small-molecule or polymer guests. Their characterization is also described, and several controlled release applications are suggested.

A number of small molecules can function as hosts for a variety of guest molecules, which may be small or polymeric, and form inclusion compounds (ICs) during their cocrystallization, both from solution and the melt. Urea (U), thiourea, perhydrotriphenylene (PHTP), cyclotriphosphazenes, and cyclodextrins (CD), for example (1-4), form ICs with a variety of guest molecules including many polymers. In polymer-ICs the host molecules form a crystalline lattice containing narrow parallel channels where highly extended guest polymer chains reside. Figure 1

132

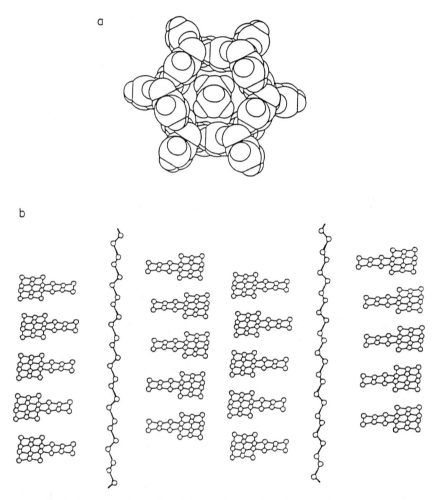

Figure 1. (a) View down the channel of the n-hexadecane-U-IC crystal (5). (b) View perpendicular to the channel of the trans-1,4-polybutadiene-PHTP-IC crystal (6).

presents views parallel and perpendicular to the channel directions for the n-hexadecane-U- and trans-1,4-polybutadine-PHTP-ICs, respectively, as determined from single crystal X-ray diffraction analyses (5,6). Note that the included guest polymers residing in different IC channels are isolated from each other and the environment by the host crystalline lattice and are highly extended because of the narrow channel cross-section (ca. 5.5Å). Similar channel structures are observed for ICs formed between these same hosts and a variety of small-molecule guests as well.

We have been forming a variety of polymer-U-ICs (7-16), and more recently polymer- and small-molecule-CD-ICs (17-19), and observing the behaviors of their included guests. In addition, we have been fabricating (20,21) molecular, small-molecule-polymer and polymer-polymer composites beginning with IC crystals. This is made possible by embedding the IC crystals into a carrier polymer phase by either a solution or melt process followed by treatment of the IC embedded carrier polymer with a solvent that is exclusive for the IC host molecules (U or CD). During the solvent treatment the IC crystals are disrupted and the included guest molecules are released and coalesce into the carrier polymer as the host molecules are removed. The present report describes this process, presents several examples of small-molecule-polymer and polymer-polymer molecular composites fabricated in this manner, and suggests some potential applications for these composites as controlled release materials.

Experimental

All polymers, U, CDs, and solvents used to fabricate the ICs and composites have been described previously (7-19). The orange, acid azodye was synthesized by Hamada (22) and contains a -CF3 label. Composites were fabricated by either a melt or solution process. Composites formed with carrier polymers whose melting temperatures are below those of the embedded ICs can be fabricated by melt pressing a film or melt extrusion of a fiber beginning with a mixture of carrier polymer and IC powders. For those composites whose carrier polymers melt above the IC crystals to be embedded, a carrier polymer solution was allowed to evaporate and IC crystals were sprinkled on top of the wet carrier polymer film. Release and coalescence of the IC guest into the carrier polymer was achieved by exposing the IC-embedded carrier polymer to a solvent for the IC host, which dissolves neither the IC-guest or the carrier polymer.

X-ray diffraction, DSC, TGA, NMR, FTIR, and water vapor permeation measurements were employed to characterize the ICs and/or the resulting molecular composites. Wide angle X-ray diffraction of powder samples were obtained at ambient conditions on a Siemens type-F X-ray diffractometer with a nickel filtered, Cu Ka radiation source (wavelength = 1.54Å). The voltage and current were set at 30kV and 20 mA, respectively. Samples were mounted on a sample holder with Scoth tape and exposed at a scan rate of $2\theta = 1°$ /min between $2\theta = 5$ and $40°$.

Perkin-Elmer Model 7 series/UNIX DSC and TGA instruments were used to observe the thermal characteristics of 5-10 mg samples heated at a rate of 10°C/min. Nitrogen was used as the purge gas for both DSC and TGA scans.

High resolution solid state C-13 NMR spectroscopy was carried out at 50.1 MHz on a Chemagnetics CMC 200S NMR spectrometer. The cross-polarized (CP), magic angle spinning (MAS), and high power (47 kHz) H-1 dipolar decoupled (DD) spectra were observed using Zirconia rotors at spinning rates of 3 to 4 kHz. The spectral width was 15 kHz stored in memory that was zero-filled to 8k before Fourier transformation.

Absorbance FTIR spectra were recorded on a Nicolet 510p spectrometer with OMNIC software at frequencies between 400 and 4000 cm^{-1}, with a resolution of 2 cm^{-1}, gain = 1, and scans = 128. Samples were thoroughly mixed with KBr and pressed into pellet form.

Rates of water permeation through film samples were obtained from weight loss measurements of pans containing water that were covered and sealed with the films (ASTM E96-80). Film thicknesses were obtained with a Thwing-Albert (Model II) thickness tester.

Results and Discussion

Figure 2 presents wide angle diffractograms of the α-CD-ICs formed with propionic acid, valeric acid, poly(ethylene oxide) (PEO), and poly(ϵ-caprolactone) (PCL) (17). Because the valeric acid-α-CD-IC is known from single crystal X-ray diffraction to adopt a channel IC structure (23,24), we conclude that PEO and PCL also adopt a channel IC structure with α-CD based on the closely similar powder diffractograms observed for the α-CD-ICs with valeric acid, PEO, and PCL. We observe in Figure 3 that the PEO-α-CD-IC is significantly more heat stable than α-CD, an observation also made for the PCL-α-CD-IC (17).

Figures 4 and 5 present the FTIR and CPMAS/DD C-13 NMR spectra of PCL-α-CD-IC and its constituents. Though not shown here, the DSC scans for PCL-α-CD-IC indicate no free PCL in the sample. However, the 1739 cm^{-1} C=O stretching band of PCL also appears in the PCL-α-CD-IC sample, thereby implying that all PCL in the sample is included in the IC crystals. This conclusion receives further confirmation in Figure 5 where carbonyl (173 ppm) and methylene (20-40 ppm) carbon resonances are observed in the CPMAS/DD C-13 NMR spectrum of PCL-α-CD-IC.

The wide angle X-ray diffractograms observed (18) for nylon-6,6-β-CD-, nylon-6-β-CD-, and nylon-6-α-CD-ICs are presented in Figure 6, where we note the diffractogram for nylon-6-α-CD is very similar to the channel-IC diffractograms observed for the PEO- and PCL-α-CD-ICs. The orange, acid azodye, whose structure is drawn in Figure 7, forms an IC with β-CD as confirmed (19) by the diffractograms of azodye-β-CD-IC and its constituents also seen there.

Low magnification micrographs of solution-cast poly (L-lactic acid) (PLLA) films embedded with PCL-U-IC before and after dipping in methanol are shown in Figure 8 (20). The PCL-U-IC domains are 50-200 μm in size, and then remain or increase from this size after the PCL chains are coalesced from the PCL-U-IC

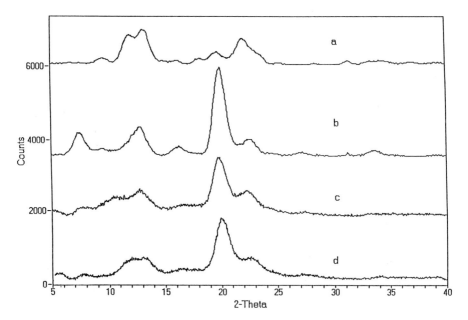

Figure 2. Wide angle X-ray diffraction of the α-CD-ICs formed with (a) propionic acid, (b) valeric acid, (c) PEO, and (d)PCL (17).

136

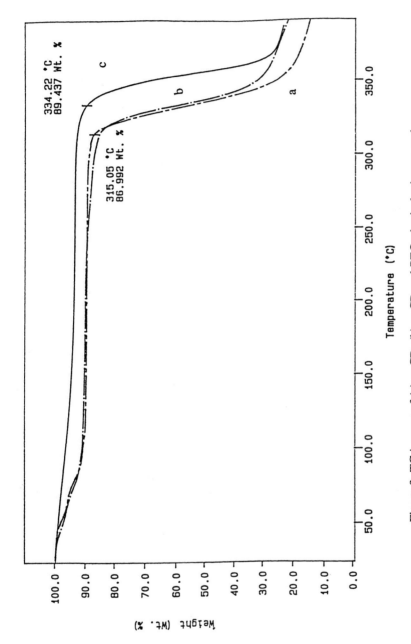

Figure 3. TGA scans of (a) α-CD, (b) α-CD and PEO physical mixture, and (c) PEO-α-CD-IC (17).

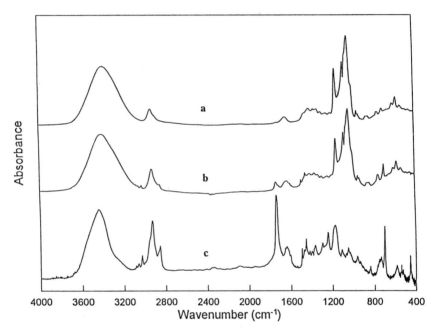

Figure 4. Fourier transform infrared spectra in the region between 400 and 4000 cm : (a) α-CD, (b) PCL-α-CD-IC, and (c) PCL (17).

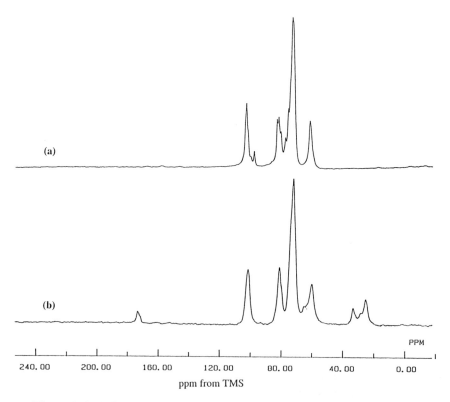

Figure 5. C-13 CPMAS/DD NMR spectra for (a) α-CD and (b) PCL-α-CD-IC.

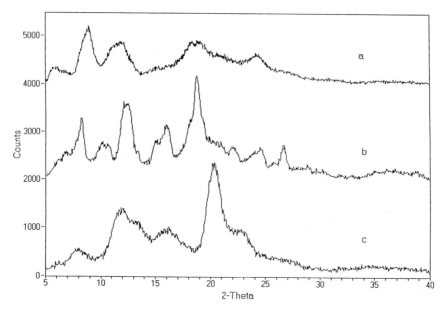

Figure 6. Wide angle X-ray diffraction of (a) nylon-66-β-CD-IC, (b) nylon-6-β-CD-IC, and (c) nylon-6-α-CD-IC (18).

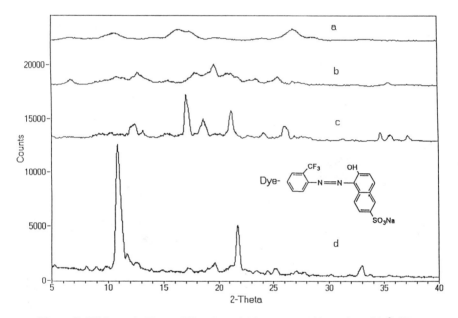

Figure 7. Wide angle X-ray diffraction of (a) orange, acid azodye, (b) β-CD, (c) orange, acid azodye and β-CD physical mixture, and (d) orange, acid azo-dye-β-CD-IC (19).

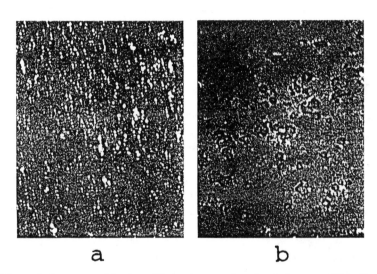

Figure 8. Low magnification (70x) micrographs of solution-cast PllA films embedded with PCL-U-IC before (a) and after (b) dipping in methanol (20).

during the methanol dip. This was not anticipated since the molar ratio of U/PCL repeat units is 4/1 or 2/1 (g/g) in PCL-U-IC (7), so we rather expected the domain sizes to decrease. Figure 9 presents the DSC scans for the PCL-U-IC embedded PLLA film before and after methanol dipping, where it is apparent that the methanol dipping has disrupted the PCL-U-IC and removed the U, because in the dipped film only the high temperature endotherm corresponding to the melting of PLLA remains. Though not shown here, CPMAS/DD C-13 NMR observation of the dipped film does exhibit PCL resonances, so the PCL chains apparently have not crystallized in the PLLA carrier film.

Table I presents a compilation of film thicknesses and moisture vapor permeabilities through PLLA solution-cast films with and without embedded U or PCL-U-IC crystals, both before and after dipping in methanol. The pure and embedded PLLA films show comparable permeabilities before the methanol dip, but after dipping only the pure and PCL-U-IC embedded films show similar water permeabilities, while the dipped U-PLLA film shows a greatly enhanced moisture transport. From these observations we conclude that methanol dipping removes U creating holes in the film which are sealed by the released PCL chains in the PCL-U-IC/PLLA dipped film.

Figure 10 presents the DSC scans of PCL fibers spun from pure PCL and from mixtures of PCL and PCL-U-IC powders (20). Each fiber was spun at 80° C, a temperature above and below the melting points of PCL and PCL-U-IC, respectively. Note that the presence of unmelted PCL-U-IC crystals seems to have induced the chain-extended crystallization of the carrier PCL chains (Tm = 63°C), while the pure PCL fiber has crystallized in the normal chain-folded morphology (Tm = 58° C). The higher temperature endotherms seen in the PCL-U-IC embedded PCL fibers are more characteristic of free U (Tm = 134° C) than PCL-U-IC (Tm = 142° C) (7), so the PCL-U-IC crystals may have been disrupted during fiber spinning. In either case, a methanol dip should remove the U to yield pure PCL fibers that may have two populations of PCL chains with different Tm's, chain-folded carrier polymer and IC-derived, extended-chain PCL (4,10,25). It would seem to be of interest to conduct careful annealing studies of these fibers in an attempt to control the types and amounts of crystalline morphologies developed.

DSC scans of a melt processed PLLA film embedded with PEO-α-CD-IC crystals before and after dipping in hot water are presented in Figure 11 (21). Note the small low temperature endotherm in the dipped film attributable to PEO chains released and coalesced from PEO-α-CD-IC. Though water is a solvent for both CD and PEO, apparently when the PEO-α-CD-IC is embedded in a PLLA film, the hot water is much more efficient in removing the CD and thus the embedded PEO chains remain trapped in the PLLA matrix.

A PEO-nylon-6 composite film sample was fabricated by embedding PEO-α-CD-IC in nylon-6 and soaking the embedded film in hot water. From the DSC scans presented in Figure 12, it is apparent that the hot water dip has released the PEO chains from their α-CD-IC crystals and permitted them to crystallize in situ in the

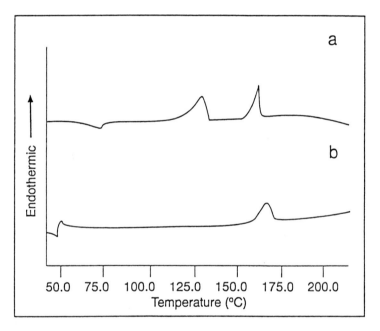

Figure 9. DSC scans of the solution cast film of PLLA embedded with PCL-U-IC before (a) and after (b) dipping in methanol (20).

Table I Comparison of Thicknesses and Permeabilities of
PLLA, PLLA-Urea, and PLLA-IC Solution-Cast Films Before and After Dipping in Methanol

Sample Name	Average Thickness (mm)	Moisture Vapor Permeability $(g/m^2/24 \ h)$
PLLA Film	0.024	173
Dipped PLLA Film	0.041	187
PLLA-Urea Film	0.180	207
Dipped PLLA-Urea Film	0.155	540
PLLA-IC Film	0.045	183
Dipped PLLA-IC Film	0.052	236

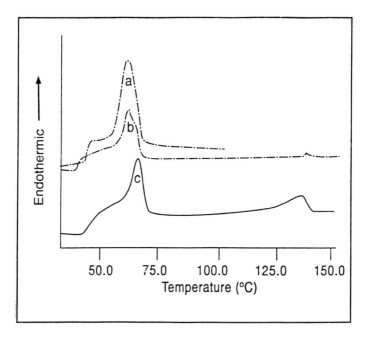

Figure 10. DSC scans of pure PCL fiber (a), PCL fiber spun with 2-3 wt% of embedded PCL-U-IC (b), and PCL fiber spun with 15-20 wt% of embedded PCL-U-IC (c) (20).

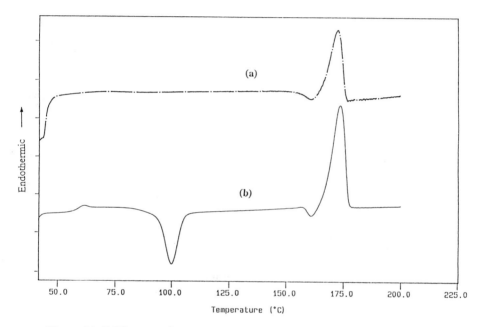

Figure 11. DSC scans of PLLA films embedded with PEO-α-CD-IC before (a) and after (b) dipping in hot water (21).

147

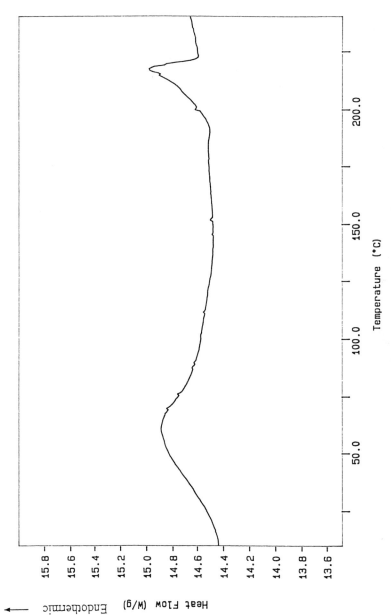

Figure 12. DSC scans of a nylon-6 film embedded with PEO-α-CD-IC before (a) and after (b) dipping in hot water (21).

nylon-6 film, which is confirmed by the presence of the low temperature endotherm due to melting of the released PEO chains in the dipped film.

We have spun nylon-6 fibers containing 10 wt% nylon-6-α-CD-IC. Currently we are measuring the moduli of these fibers before and after dipping in hot water, which should release the IC-nylon-6 chains. Comparisons will be made with identically spun pure nylon-6 fibers, and we plan to treat each of the fibers with identical annealing and fiber-drawing regimens to learn whether or not the strengths of nylon-6 fibers might be improved by their IC-assisted fabrication (21).

A nylon-6 film has been selectively embedded with a small portion of the orange, acid azodye-β-CD-IC in a melt process (21). The azodye-β-CD-IC crystals have been confined to a small area in the center of the film. We hope to dip the dye-embedded nylon-6 film in hot water and observe the migration of the dye throughout the nylon-6 film after it is released from its β-CD-IC crystals. These observations of inside/out dyeing should permit an estimate of the dye diffusion coefficient which is free from the usual complications encountered in dye diffusion estimates obtained from the outside/in dyeing of fibers in a dye bath.

Having demonstrated the fabrication of molecular composites formed between small-molecules and polymers and between different polymers, beginning with ICs formed with U and CD hosts and guest small-molecules and polymers and followed by the subsequent release and coalescence of the IC guests into a carrier polymer phase, we close this contribution by suggesting several potential controlled release applications of these IC-fabricated composites. Production of films containing islands of an IC-derived polymer in a sea of carrier polymer, may lead to membranes whose permeabilities to gases and liquids differ among the phase separated regions. If the IC-derived polymer islands are more permeable or may be removed from the matrix carrier polymer of the composite film by dissolution, then membranes with porous regions or pores may be obtained, thus leading to membranes with tailorable permeabilities.

It has been observed (26) that PCL and racemic poly(L,D-lactic acid) (PLDLA) respond quite differently in biodegradation and drug-release studies. PLDLA films decompose at a rate three times that observed for PCL, while the release of steroids from PCL films is nearly two orders of magnitude faster than from PLDLA films. Clearly drug-release from polymer films is not directly tied to their over all macroscopic degradation. An optimal combination of both characteristics might be achieved by embedding PCL-urea-IC crystals in a PLDLA film, followed by the release and coalescence of the PCL chains from their U-IC crystals into the carrier PLDLA. In this manner a film composed of a readily degradable matrix (PLDLA) embedded with PCL domains, from which facile drug-release occurs, can be achieved. Such a composite film would serve as an effective drug-releasing implant, with the added benefit that the biologically inactive film portion of the implant (the drug container) would also rapidly degrade and be absorbed by the body. This suggested example is but one of many possible to visualize based on the unique phase separated morphologies that are possible to create with polymers embedded, released, and coalesced from their IC crystals into a second polymer phase which serves as a carrier matrix.

As illustrated by the orange, acid azodye-β-CD-IC mentioned previously, we may fabricate small-molecule-polymer composites beginning with small-molecule-CD-Cs. For example, we have embedded the orange, acid azodye-β-CD-IC in a nylon-6 film, and we anticipate that exposure of the film to hot water will disrupt the dye-CD-IC crystals, remove the host β-CD, and release the guest dye, thereby dyeing the nylon-6 film from the inside out. ICs formed between drugs, pesticides, and other biologically active small-molecules and hosts such as CD should be easily embedded or encapsulated in carrier polymer phases for eventual release of the active IC guest. The active IC guest may be coalesced into the carrier polymer phase before use or the active guest may be retained in the IC crystals for eventual release into the carrier polymer when exposed to the appropriate environments. Utilization of CD or certain CD derivatives to form the active guest-CD-IC should permit control of the release of the active IC-guest through control of the solubilities of the IC-hosts. In this way it may be possible to not only control the release of the embedded, active IC-guest, but its release may also be targeted to particular environments based on the solubility of the IC-host.

We have successfully spun fibers embedded with both small-molecule-and polymer-ICs. This affords the opportunity to create composite fibers when the IC guests are released into the carrier polymer fiber. For example, if we were to embed PLLA fibers with an antibiotic-CD-IC, the antibiotic-PLLA composite fiber obtained by releasing antibiotic from the CD-IC and removing the CD might be usefully employed as suture material containing its own internal antibiotic for preventing infection in the sewn incision.

We have speculated on the potential utility of small molecule and polymer-polymer molecular composites fabricated form their embedded ICs in the hope that other researchers will also recognize potential applications for these unique composites.

Literature Cited

1. L. C. Fetterly, "Non-Stoichiometric Compounds," L. Mandelcorn, Ed., Academic Press, New York, 1964, Chap. 8.
2. K. Takemoto and N. Sonoda, "Inclusion Compounds," J. Attwood, J. Davies, and D. MacNicol, Eds., Academic Press, London, Vol. 2, 1984, Chap. 2.
3. M. Farina, "Inclusion Compounds," J. Attwood, J. Davies, and D. MacNicol, Eds., Academic Press, London, Vol. 2, 1984, Chap. 10.
4. G. Silvestro and P. Sozzani, Comprehensive Polymer Science," G. Eastman et al., Eds., Pergamon Press, Oxford, Vol. 2, 1988, Chap. 18.
5. A. E. Smith, Acta Crystallogr., 5, 224 (1952).
6. A. Colombo and G. Allegra, Macromolecules, 4, 579 (1971).
7. C. Choi, D.D. Davis, and A. E. Tonelli, Macromolecules, 26, 1468 (1993).
8. C. Howe, S. Sankar, and A. E. Tonelli, Polymer, 34, 2674 (1993).
9. N. Vasanthan, A. E. Tonelli, and S. Nojima, Macromolecules, 27, 720 (1994).

10. N. Vasanthan, I. D. Shin, and A. E. Tonelli, Macromolecules, 27, 6515 (1994).

11. C. Howe, N. Vasanthan, C. MacClamrock, S. Sankar, I. D. Shin, I. Simonsen, and A. E. Tonelli, Macromolecules, 27, 7433 (1994).

12. N. Vasanthan and A. E. Tonelli, Polymer, 36, 4887 (1995).

13. N. Vasanthan, I. D. Shin, and A. E. Tonelli, J. Polym. Sci., Polym. Phys. Ed., 33, 1385 (1995).

14. N. Vasanthan, I. D. Shin, and A. E. Tonelli, Macromolecules, 29, 263 (1996).

15. P. Eaton, N. Vasanthan, I. D. Shin, and A. E. Tonelli, Macromolecules, 29, 2531 (1996).

16. N. Vasanthan, I. D. Shin, L. Huang, S. Nojima, and A. E. Tonelli, Macromolecules, 30, 3014 (1997).

17. L. Huang, E. Allen, and A. E. Tonelli, Polymer, 38, XXXX (1997).

18. L. Huang, E. Allen, and A, E. Tonelli, J. Polym. Sci., Polym. Phys., Ed., 35, XXXX(1997).

19. Y. Song, L.Huang, K. Hamada, and A. E. Tonelli, Macromolecules, 31, XXXX (1998).

20. L. Huang, N. Vasanthan, and A. E. Tonelli, J. Appl. Polym. Sci., 64, 281 (1997).

21. L. Huang, E. Allen, and A. E. Tonelli, J. Appl. Polym. Sci., 64 (1998).

22. K. Hamada, Synthesis of orange, acid azodye paper.

23. K. Takao and T. Kuge, Agr. Biol. Chem., 34, 17787 (1970).

24. R. K. Mc Mullan, W. Saenger, J. Fayos, and D. Mootz, Carbohydr. Res., 31, 37 (1973).

25. Y. Chatani and S. Kuwata, Macromolecules, 8 12 (1975).

26. A. Schindler, R. Jeffcoat, G. L. Kimmel, C. G. Pitt,M. E. Wall, and R. Zweidinger, "Contemporary Topics in Polymer Science," E. M. Pearce and J. R. Schaefgen, Eds., Vol. 2, Plenum, New York, 1975, p. 251.

Chapter 11

Liposomal Delivery Systems
for Phototherapeutic Agents

P. W. Taylor

Novartis Pharmaceuticals, Wimblehurst Road,
Horsham, West Sussex RH12 4AB, United Kingdom

Numerous attempts have been made to improve the therapeutic potential of a wide range of pharmacologically active agents by enhancement of their efficacy and reduction of their toxicity. The unique properties of liposomes make them particularly promising vehicles for modifying drug pharmacokinetics in a way that allows their accumulation at disease sites and reduces their distribution to sensitive, non-target tissues. Liposomes are stable, spherical assemblies in which an aqueous volume is entirely enclosed by a membrane, usually composed of a bilayer of phospholipid molecules. They may be composed of one or more concentric lipid bilayers and their size and lamellarity is determined to a large extent by the method used for their preparation and by the nature of the constituent lipids. The presence of an aqueous compartment, which distinguishes them from other particulate carrier systems such as oil-in-water emulsions, can be used to accomodate hydrophilic drugs, whereas hydrophobic drug molecules can be incorporated into the lipid bilayers, often in therapeutically meaningful quantities. Thus, liposomes have been used to entrap a wide variety of drugs that include anticancer (1), antifungal (2), antibacterial (3) and anti-inflammatory (4) agents. Traditionally, the membrane components of liposomes have been phospholipids, particularly phosphatidyl cholines, partly because they are the natural building blocks of biological membranes and partly because the common phospholipids are lamella-forming lipids under all conditions and form stable structures without the need for other components. However, incorporation of cholesterol into phospholipid bilayers has facilitated the design of vehicles with altered rates of drug leakage and residence time in the circulatory system. Recently, bilayer membrane vesicles have been constructed using single-chain amphiphiles or non-ionic surfactants which have properties similar or complimentary to conventional liposomes (5).

It has recently been pointed out (6) that the development of liposomal drug delivery systems has progressed to the point where methodologies for efficient encapsulation of a variety of drug molecules in high drug:lipid ratios using biocompatible lipids, extension of their blood circulation time for periods of 24 h or more and deposition at target sites by either active or passive targeting mechanisms are readily achievable. Indeed, the last few years have seen the introduction of successful commercial products based on liposomal formulations that facilitate the efficaceous delivery of the antifungal agent amphotericin B (7) and the antiproliferatives doxorubicin and daunorubicin (8,9) and there are liposomal formulations of other conventional drugs, protein therapeutics and gene and antisense oligonucleotides that are currently in preclinical or early clinical development (9).There are, however, a number of factors that limit the usefulness of liposomes as drug carriers for use in man. Many liposomal preparations are highly interactive with components of blood and other biological fluids and this may result in premature leakage of drug payload within the aqueous compartment or the release of lipophilic drug molecules and their subsequent partitioning into lipidic structures unrelated to the pharmacological target. Although it is envisioned that the next generation of liposomal pharmaceuticals will carry surface-associated cell targeting information (9), site-selective uptake is at present limited to components of the mononuclear phagocyte system, such as the liver and spleen, and to anatomical regions possessing a leaky vasculature. Active targeting mediated through surface-bound recognition features such as monoclonal antibodies, lectins and receptor ligands raise issues of immunogenicity, manufacturing complexity and cost; advantages with respect to efficacy and reduced toxicity remain as yet unproven. Likewise, the formidable barriers to selective delivery to intracellular targets are yet to be overcome.

In addition to their capacity to advantageously alter drug pharmacokinetic and biodistribution profiles, liposomes may also fulfill the basic function of providing a suitable vehicle for the systemic delivery of "difficult to administer" drugs. They may also protect drugs that are rapidly degraded *in vivo*, such as cytosine arabinoside, and provide sustained release and depot systems for systemic and topical delivery. These, and other, issues will be encountered during the preclinical development of liposomally formulated therapeutic agents and it is proposed, in the rest of this article, to illustrate this point by reference to attempts to develop liposomal dosage forms of phototherapeutic agents for use in the photodynamic therapy of hyperproliferative conditions.

The Concept of Photodynamic Therapy

Photodynamic therapy (PDT) is a treatment modality that takes advantage of the fact that photosensitive tetrapyrolic molecules, such as porphyrins, phthalocyanines and bacteriochlorins, are able to accumulate in and be selectively retained by abnormal or hyperproliferative cells and tissues.

Subsequent photoactivation of these molecules in the presence of oxygen may lead to destruction of the target tissue, while sparing surrounding normal tissue, as a result of singlet oxygen mediated peroxidative damage (10,11). Although at present PDT is considered predominantly a modality for the ablation of solid tumors, there is growing interest in its potential for the treatment of a variety oncological, cardiovascular, dermatological, ophthalmic and infectious indications (12-15).

The concept of photochemical sensitization of target cell populations was formulated several decades ago and the first full clinical report of PDT appeared in 1976. In the last ten years there has been rapid progress to a point where products are on the market, with more likely to follow in the near future. Currently, photosensitizers are administered predominantly by the systemic route, but there is growing interest in the use of topical formulations (16). Activation of the drug in tissue involves the application of light generated by a laser and delivered selectively through a fibre optic device; optical technology is evolving rapidly as the specific requirements for clinically effective PDT become clearer (17). The sensitizer used most frequently for PDT in the clinic is the commercially available preparation Photofrin II®, a mixture of non-metallic oligomeric porphyrins linked primarily through ether bonds (18). Although effective, particularly with respect to ablation of superficial bladder cancer, obstructive endometrial tumors, esophageal tumors and primary and metastatic skin tumors, it produces prolonged skin photosensitvity and has other drawbacks that are related to its less than ideal chemical, photochemical and biological properties. Consequently, there is currently much interest in developing second generation photosensitizers such as benzoporphyrin derivative (11) that have a more attractive profile. Ideally, photosenstizers should be relatively easy and inexpensive to synthesise, have no toxic side effects in the absence of light, be photostable and have a high absorption coefficent for photoactivating light, which should be at wavelengths (650-750nm) giving good tissue penetration and for which suitable light sources exist. In addition, they should lack mutagenic potential, generate a high singlet oxygen yield following activation and be selectively and preferentially retained in the target tissue with rapid elimination from normal tissue. Very few currently available photosensitizers possess all these characteristics (19).

The rate and extent of uptake into target tissue is strongly influenced by the physicochemical properties of the photosenstizer and by the mode of delivery of the drug. For example, hydrophilic molecules, such as tri- and tetrasulfonated derivatives of porphyrins and phthalocyanines, tend to bind in a noncovalent fashion to plasma proteins, particularly albumin and globulins, and subsequently localise in the vascular stroma of target tumors (20,21). In situ photoactivation causes damage to the tumor microvasculature, resulting in vascular stasis, anoxia and tumor necrosis and the process may be augmented by photosensitizer binding to components of the nonvascular stroma, such as collagen and elastin. In contrast, lipophilic photosensitizers associate preferentially with plasma lipoproteins following administration into the bloodstream (*21*). Studies with *in vitro* systems and with animal models demonstrate unequivocally that lipophilic photosensitizers bind strongly to the three main classes of lipoprotein particles found in man, *viz* high- (HDL), low- (LDL) and very low-density lipoproteins (VLDL) and binding occurs proportionately to the relative abundance of each lipoprotein class in plasma (*22-25*). Lipoprotein-bound photosensitizers have a tendency to be taken up by neoplastic cells within a tumor and there is some indication that this may result from the capacity of hyperproliferating cells to internalise LDL (*26,27*). Following cell binding, lipophilic photosensitizers associate predominantly with cell membranes (*28,29*) and photoactivation consequently results in direct killing of the target cell (*30*).

Lipid-based delivery systems for lipophilic photosensitizers

The flat aromatic macrocycle confers on many porphyrins and their analogues a strong tendency to undergo aggregation in neutral aqueous solutions. The photophysical and biological properties of lipophilic photosensitizers are adversely affected by aggregation (*31*) and it is therefore imperative to ensure that preparations for use in PDT be maintained, as much as is possible, in the monomeric state. Early studies, reviewed by Jori and Reddi (*21*), determined that a number of lipophilic photosensitizers could be incorporated in a stable fashion into liposomes and, in addition to providing a vehicle to facilitate the administration of the drugs in a biologically active form, appeared to improve the rate and extent of accumulation of the dyes in tumor target tissue. Due to their high degree of hydrophobicity, these molecules incorporate into the liposomal phospholipid bilayer and after transfer to cellular targets are buried in apolar lipid regions of the tumor cells, which hinders their clearance from the target site.

Jori and coworkers have systematically examined the capacity of a variety of liposomal formulations to deliver lipophilic photosensitizers, notably zinc(II)-phthalocyanine (Zn-Pc), to implanted tumor targets; in addition, they have provided strong evidence of a role for plasma lipoproteins in the localisation process. *In vitro* complexation of photosensitizer with LDL led to greater tumor tissue accumulation of the dye in comparison to aqueous suspensions (*32*) and similar tumor localisation patterns could be achieved when Zn-Pc was formulated with small unilamellar vesicles composed of dipalmitoyl phosphatidylcholine (DPPC;*33*). Low doses (70-350 μg kg^{-1}) of DPPC-delivered Zn-Pc were suffcient to induce a necrotic response in an implanted MS-2 fibrosarcoma. The degree of dye retention by the tumor was such that PDT produced a response even if performed up to 70 h after drug administration (*33*). Zn-Pc administered in DPPC liposomes was transported in the bloodstream by plasma lipoproteins and a maximum concentration of Zn-Pc in the tumor was observed after 18-24 h (*23*). Incubation of Zn-Pc incorporated into DPPC, distearoyl phosphatidylcholine (DSPC) or dimiristoyl phosphatidylcholine (DMPC) liposomes with human serum resulted in binding of the photosensitizer to VLDL, LDL and HDL; incorporation of cholesterol into DPPC liposomes increased the transfer to LDL. A similar association with lipoproteins was observed after serum was recovered from rabbits administered the various liposomal formulations (*34*). Liposomal Zn-Pc produced tumor necrosis in transplanted MS-2 fibrosarcomas following red light irradiation and early photodamage to mitochondria and rough endoplasmic reticulum of malignant cells was evident before any changes in capillaries supplying the tumor were apparent (*35*). Thus, damage mediated by photoactivation of this lipophilic molecule is predominantly directed against cells of the tumor mass. Similar pharmacokinetic and phototherapeutic data has been obtained using liposomal formulations and emulsions of tin-etiopurpurin (SnET2) and a siloxynaphthalocyanine (*36,37*).

The second generation photosensitizer benzoporphyrin derivative monoacid ring A (BPD), a hydrophobic molecular equimixture of two regioisomers, has been incorporated into a liposomal formulation and is currently undergoing extensive clinical profiling in a number of indications, including age-related macular degeneration of the retina, a leading cause of blindness in the elderly (*12*). BPD loaded into DMPC-containing unilamellar vesicles accumulated in tumor tissue in MI rhabdomyosarcoma-bearing mice to a greater extent than DMS/PBS-solubilised BPD and this resulted in a superior phototherapeutic effect following drug activation (*38*). Liposomally delivered drug appeared to enter non-target tissues such as skin, lung and fat more rapidly and to be cleared from these tissues more rapidly, indicating that lipsomal formulation may reduce drug-induced skin photosensitivity resulting from accumulation and persistence of photosensitizer in this organ. Liposomal BPD associated with the three main classes of human plasma lipoproteins when examined in *in vitro* distribution assays (*38*). Clinical studies are also being conducted with lipid emulsions of SnET2 in order to determine efficacy in patients with cutaneous malignancies and Karposi's sarcoma (*39*).

Development of liposomal Zn-Pc

The chemical, photochemical and biological properties of unsubstituted Zn-Pc make it an attractive candidate for clinical evaluation (*40*). It can be produced in a cost-effective manner by a one-step synthesis from benzodinitrile followed by extensive purification (*41*) and exhibits a high degree of physical and chemical stability. It has a high extinction coefficient, a long triplet lifetime, a high singlet oxygen yield and good fluorescence properties. Its absorption maximum of 671 nm (Figure 1) should ensure minimal skin photosensitization problems and at this wavelength light penetrates tissues to a depth of about 5 mm and hence causes a deeper tissue/tumor necrosis than photosensitizers such as Photofrin II[®](*42*).

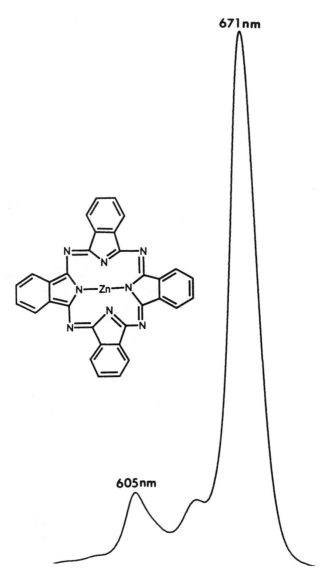

Figure 1. Structure and action spectrum of Zn-Pc

Its extremely poor solubility in aqueous milieu and all but a handful of organic solvents (43) warrants the use of lipid-based formulations for parenteral administration. Jori and coworkers had demonstrated (44) that, over a certain concentration range, Zn-Pc could be incorporated into unilamellar vesicles of DPPC predominantly in the monmeric (non-aggregated) state, an essential prerequisite for effective photodynamic therapy. Systematic appraisal of the criteria for pharmacological efficacy and large-scale manufacturing suitability resulted in the selection of a Zn-Pc dosage form comprising small unilamellar vesicles of 1-palmitoyl, 2-oleoyl phosphatidylcholine (POPC) and 1,2-dioleoyl phosphatidylserine (OOPS) in Zn-Pc:POPC:OOPS w/v ratios of 1:90:10 (45). These synthetic phospholipids were utilised due to their consistent composition with respect to fatty acid content, their stability to oxidation in comparison to polyunsaturated analogs from natural sources and to their availibility in bulk. The manufacturing procedure for this formulation (designated CGP 55 847) was essentially a solvent dilution method that yielded vesicles with diameters in the 65-100 nm range in which Zn-Pc molecules had been incorporated into the phospholipid bilayer predominantly in the monomeric state (45). In order to ensure that the procedure was suitable for the large-scale production of liposomes for systemic administration to man, N-methylpyrrolidone (NMP), a biodegradable, non-toxic, water-miscible organic solvent (46), was used to solubilise Zn-Pc. As phospholipids exhibit low solubility in NMP, the non-toxic, water-miscible solvent t-butyl alcohol was used in order to solubilise POPC and OOPS. After mixing of these two solutions, the components were rapidly diluted in an excess of aqueous phase using a dynamic mixing device (45). By careful selection of electrolyte/non-electrolyte composition of the aqueous phase and the conditions under which the components were mixed, Zn-Pc-containing small liposomes were formed in a reproducible fashion without the need for a further size reduction step. After concentration, organic solvents were removed by cross-flow filtration and the liposomes were sterile filtered. Lyophilisation in the presence of cryoprotectant yielded stable preparations (vials containing 0.4 mg Zn-Pc and 40 mg phospholipid) that could be reconstituted with pyrogen-free water prior to administration. There was no change in the composition and size distribution of liposomes that were reconstituted after storage for one year at various temperatures and Zn-Pc was found to be completely retained in the phospholipid bilayer after this period of time (41). Liposomes produced by this technique were impressively homogeneous with respect to particle size distribution (45).

Schieweck and coworkers have undertaken extensive pharmacokinetic and phototherapeutic investigations of CGP 55 847 (41,47). The uptake and elimination of Zn-Pc in tissues and organs of Meth A sarcoma-bearing mice after intravenous administration of 0.125 or 0.25 mg/kg liposomal Zn-Pc were dose related. The liposomally-delivered photosensitizer was cleared from the bloodstream with a half-life of about 6 h but a small proportion of the injected dose of Zn-Pc persited for a number of days. In healthy mice, there was rapid uptake of Zn-Pc by the liver followed by relatively fast elimination. Again, minimal amounts of Zn-Pc persisted in this organ (*41*). Systemic administration of phototherapeutic amounts of CGP 55 847 to tumor-bearing mice resulted in maximum concentrations of Zn-Pc in the tumor tissue after 48 h and remaining relatively constant up to 168 h. Significantly higher photosensitizer concentrations were found in the implanted tumors in comparison to peritumoral tissue (Table I), indicating a high degree of tissue selectivity of liposomally-delivered Zn-Pc. Zn-Pc concentrations in implanted tumors were in the range 250400 ng/g tissue between 48 and 168 h following liposomal administration (41). Tumor selectivity was also evident when implanted Ehrlich carcinomas from CGP 55 847-dosed tumor-bearing mice were resected and sections examined histologically; Zn-Pc visualised by fluorescence microscopy was clearly associated with the membranes and cytosol of carcinoma cells and not to any extent with proximal muscle fibers (*47*). The importance of preserving the monomeric state of the photosensitizer was illustrated by the fact that induced Zn-Pc aggregation in the liposomal bilayer resulted in an increased rate of clearance from the bloodstream, a lowering of the maximum photosensitizer concentration in the target tumor tissue and increase in concentration and persistence of the dye in the liver (*48*). Unlike studies with other photosensitizers, it was demonstrated that photosensitizer concentrations in the tumor were comparable to those found in the liver, the organ with the greatest capacity to accumulate photosensitizer from the circulation. The rate of clearance of Zn-Pc from tumor tissue was much lower than from the major organs, providing a suitable time frame for photoactivation of the drug, which could include multiple light applications (*41*).

Table I. Tumor-to-tissue ratios of liposomally-delivered Zn-Pc administered i.v. to Meth A sarcoma-bearing BALB/C mice. Adapted from Schieweck *et al* (*41*).

Tumor/ Tissue	Zn-Pc	Hours after intravenous injection of CGP 55 847						
	mg/kg	3	6	24	48	72	96	168
Tumor/ Muscle	0.250	4.1:1	5.5:1	7.0:1	10.0:1	10.2:1	13.6:1	14.5:1
	0.125	4.4:1	6.1:1	9.8:1	8.2:1	9.5:1	11.6:1	13.7:1
Tumor/ Skin	0.250	2.7:1	3.6:1	2.8:1	3.5:1	3.0:1	3.2:1	3.0:1
	0.125	2.4:1	2.2:1	3.6:1	3.1:1	3.0:1	3.8:1	3.8:1
Tumor/ Liver	0.250	0.16:1	0.19:1	0.57:1	1.06:1	1.29:1	1.47:1	1.58:1
	0.125	0.15:1	0.20:1	0.65:1	1.05:1	1.27:1	1.71:1	1.84:1

Tumor-bearing mice were also subjected to PDT, following intravenous administration of CGP 55 847, using red light of 671 nm wavelength generated by an argon-dye laser (41). Doses as low as 0.063 mg/kg of Zn-Pc caused tumor necrosis and subsequent eradication of tumor burden when Meth A sarcomas implanted in mice were irradiated with moderate doses of light (~300-350 Joules/cm^2). A simlar total curative effect could be achieved with only 9 Joules/cm^2 when higher doses (0.250 or 0.375 mg/kg Zn-Pc) of CGP 55 847 were administered.It was clearly established that the key parameters for successful PDT were drug dose and total light dose and the interrelationship between the two has been elegantly elucidated by Schieweck's group (41). With certain doses, sufficient Zn-Pc was retained in the tumor to facilitate light-induced necrosis one week after a single administration of the liposomal formulation. Tumor necrosis leading to cure was also found after PDT in rats implanted with a mammary carcinoma, in nude mice with a human bladder carcinoma and in mice with chemically-induced skin carcinoma. In addition, the phototoxic potential of CGP 55 847 in hairless mice was low (41)

Delivery of liposomal Zn-Pc to tumor targets

It is clear that lipophilic photosensitizers have the capacity to partition into plasma lipoproteins and that the extent to which this occurs may be increased by entrapment within certain liposomal formulations (21). Continuously replicating cancer cells require relatively large amounts of cholesterol for membranogenesis and readily take up LDL (49,50). As this process is often receptor-mediated, malignant cells frequently express elevated numbers of LDL receptors in comparison to non-malignant cells (51,52). Thus, the receptor-mediated endocytosis of LDL represents a mechanism for the selective internalisation of lipophilic photosensitizers by tumor cells with the advantage that photosensitizers located intracellularly are more efficient in cell inactivation following irradiation than extracellular molecules (53). Although direct insertion of photosensitizers into LDL obtained from plasma samples is feasible (21) there are severe technical limitations to the development of such a strategy for the routine administration of drugs. One way forward would be to develop recombinant or synthetic drug-loaded particles for selective delivery via receptors involved in cholesterol metabolism (54) or, more cost-effectively, to exploit interactions between liposomes and plasma lipoproteins.

We have obtained clear evidence that CGP 55 847 interacts vigorously with plasma lipoproteins. Short incubation (<2 min) of highly purified human LDL with increasing concentrations of liposomes led to a progressive increase in the net negative charge of LDL as determined by agarose gel electrophoresis and both Zn-Pc and liposomal phospholipid were incorporated into the modified particles, albeit in differing amounts (55). There was an increase in the molecular mass of liposome-modified LDL and immunoaffinity chromatography provided evidence that apoprotein B epitopes on modified LDL were unable to bind effectively to anti-apoprotein B antibody. Thus, the incorporation of components of CGP 55 847 into LDL resulted in some modification of the properties of the lipoprotein particles but provided evidence that LDL may be able to function as the primary vehicle for selective delivery of liposomally-formulated to cells expressing increased numbers of LDL receptors. Polo *et al* (*56*) confirmed the rapid and vigorous interaction between purified LDL and CGP 55 847 and showed that about 60 Zn-Pc molecules can be incorporated into each LDL particle. Transfer of photosensitizer from POPC:OOPS liposomes occurred within a few minutes in their system, whereas transfer of Zn-Pc from DPPC to LDL was found to proceed much more slowly, almost certainly indicating that a liposomal bilayer in the fluid state at 37oC is an prerequisite for rapid interaction and transfer. Thus, the composition of the liposomal vehicle plays a major role in determining the rate and extent of photosensitizer incorporation.

Liposomally-delivered Zn-Pc can also incorporate into lipoprotein particles other than LDL. When CGP 55 847 was incubated with mixtures of highly purified human HDL, LDL and VLDL (combined in amounts reflecting their natural abundance in plasma), Zn-Pc was found to be associated with all three lipoprotein classes after a very short incubation period (*57*). The bulk of Zn-Pc transferred from liposomes to lipoproteins with very little cotransfer of phospholipid. When CGP 55 847 was incubated in pooled plasma, up to 85% of Zn-Pc and about 30% of phospholipid were recovered with LDL and HDL, indicating a potential role for plasma lipid transfer proteins in the incorporation of phospholipid into lipoproteins. It was also found that the buoyant density of Zn-Pc liposomes increased in a dose-dependent manner during incubation with human plasma and opsonins, including fragments of complement component C3, were found in association with liposomes. Thus, CGP 55 847, like other liposomes with bilayers in the fluid phase at 37°C, are subject to modification by plasma proteins (*58*) in addition to their interaction with lipoproteins. As the *in vitro* data discussed immediately above would appear to suggest, LDL particles carrying liposomally-delivered Zn-Pc bind to the LDL receptor with reduced affinity (*29*), although this may not necessarily be the case with LDL containing lower amounts of Zn-Pc. Liposomal Zn-Pc is taken up by LDL receptor-expressing HepG2 cells in culture to a greater extent in the presence of LDL than in its absence (*29*), providing support at the cellular level for the contention that the LDL uptake pathway contributes to the localization of Zn-Pc in hyperproliferative tissue. It should be borne in mind that there is wide variation in the lipoprotein content of plasma from and within different species (*59*), with particularly high variability in man due to genetic, cultural and lifestyle factors. Animal studies of the distribution of photosensitizers amongst the various classes of lipoprotein particles have generally been performed in species with a significantly different lipoprotein profile in comparison to man and, indeed, there is some evidence to suggest that variations in plasma lipid composition can have a small but significant effect on the biodistribution of Zn-Pc in rodent models (*60*). These factors may prove to be of importance as PDT programs move from the preclinical into the clinical phases of evaluation.

Conclusions

Liposomes have proven to be versatile vehicles for the selective delivery of photodynamic agents, at least in the preclinical phase. They provide a suitable means by which lipophilic photosensitizers, such as Zn-Pc, can be systemically administered in the monomeric state that is necessary for effective PDT. Although vesicles with bilayers in the fluid phase at physiological temperatures are rapidly modified in the bloodstream, they facilitate the transfer of drug to lipoproteins such as LDL that may be selectively accumulated by hyperproliferating target tissues. These advantages are seen with CGP 55 847, a preparation of small unilamellar vesicles consisting of the synthetic lipids POPC and OOPS into which unsubstituted Zn-Pc has been intercalated. The excellent physical, chemical, photochemical and biological properties of Zn-Pc in combination with the high degree of tumor specificity obtained through liposomal encapsulation of the drug make this formulation an attractive candidate for clinical development. A reproducible manufacturing process allows the preparation of homogeneous batches of lyophilized liposomal Zn-Pc with a long shelf life. Liposomally-delivered Zn-Pc has an excellent pharmacokinetic profile, high uptake and retention in implanted tumors and a high selectivity for the target tissue.

Literature cited

1. McCullough, H. N.; Juliano, R. L. *J. Natl. Cancer Inst.* **1979,** *63*, 727-731.
2. Lopez-Berenstein, G.; Bodey, G. P.; Fainstein, V.; Keating, M.; Frankel, L. S.; Zeluff, B.; Gentry, L.; Mehta, K. *Arch. Intern. Med.* **1989,** *149*, 2533-2536.
3. Schreier, H.; McNichol, K. J.; Ausborn, M.; Soucy, D. M.; Derendorf, H.; Stecenko, A. A.; Gonzalez-Rothi, R. J. *Int. J. Pharm.* **1992,** *87*, 183-193.
4. Naeff, R.; Pliska, V.; Weder, H. G. *J. Microencapsulation* **1990,** *7*, 95-103.
5. Bouwstra, J. A.; Junginger, H. E. In *Liposomes, New Systems and New Trends in their Applications;* Puisieux, F.; Couvreur, P.; Delattre, J.; Devissaguet, J-P., Eds.; Editions de Santé: Paris, France, 1995; pp 99-121.
6. Gerasimov, O. V.; Rui, Y.; Thompson, D. H. In *Vesicles;* Rosoff, M., Ed.; Marcel Dekker: New York, 1996; pp 679-746.
7. Janknegt, R.; de Marie, S.; Bakker-Woudenberg, I. A. J. M.; Crommelin, D. J. A. *Clin. Pharmacokinet.* **1992,** *23*, 279-291.
8. Barenholz, Y.; Amselem, S.; Goren, D.; Cohen, R.; Gelvan, D; Samuni, A.; Golden, E. B.; Gabizon, A. *Medicinal Res. Rev.* **1993,** *13*, 449-491.
9. Chonn, A.; Cullis, P. R. *Curr. Opinion Biotechnol.* **1995,** *6*, 698-708.

10. Pass, H. I. *J. Natl. Cancer Inst.* **1993**, *85*, 443-456.
11. Dolphin, D. *Can. J. Chem.* **1994**, *72*, 1005-1013.
12. Levy, J. G. *TIB Tech.* **1995**, *13*, 14-18.
13. North, J.; Neyndorff, H.; Levy, J. G. *J. Photochem. Photobiol. B: Biol.* **1993**, *17*, 99-108.
14. Visonà, A.; Jori, G. *Atherosclerosis* **1993**, *100*, 213-222.
15. Koderhold, G.; Jindra, R.; Koren, H.; Alth, G; Schenk, G. *J. Photochem. Photobiol. B: Biol.* **1996**, *36*, 221-223.
16. Lui, H.; Anderson, R. R. *Arch. Dermatol.* **1992**, *128*, 1631-1636.
17. Unger, M. *SPIE Proc.* **1995**, *2371*, 16-21.
18. Dougherty, T. J.; Marcus, S. L. *Eur. J. Cancer* **1992**, *28A*, 1734-1742.
19. Boyle, R. W.; Dolphin, D. *Photochem. Photobiol.* **1996**, *64*, 469-485.
20. Kessel, D.; Thompson, P.; Saatio, K.; Nantwi, K. D. *Photochem. Photobiol.* **1987**, *45*, 787-790.
21. Jori, G.; Reddi, E. In *Photodynamic Therapy of Neoplastic Disease;* Kessel, D., Ed.; CRC Press: Boca Raton, Florida, 1990; pp 117-130.
22. Kongshaug, M.; Moan, J.; Cheng, L. S.; Garbo, G. M.; Kolboe, S.; Morgan, A. R.; Rimington, C. *Int. J. Biochem.* **1993**, *25*, 739-760.
23. Reddi, E.; Lo Castro, G.; Biolo, R.; Jori, G. *Br. J. Cancer* **1987**, *56*, 597-600.
24. Polo, L.; Reddi, E.; Garbo, G. M.; Morgan, A. R.; Jori, G. *Cancer Lett.* **1992**, *66*, 217-223.
25. Allison, B. A.; Pritchard, P. H.; Richter, A. M.; Levy, J. G. *Photochem. Photobiol.* **1990**, *52*, 501-507.
26. Kessel, D. *Cancer Lett.* **1986**, *33*, 183-188.
27. Jori, G.; Reddi, E. *Int. J. Biochem.* **1993**, *25*, 1369-1375.
28. Peng, Q.; Moan, J.; Farrants, G.; Danielsen, H. E.; Rimington, C. *Cancer Lett.* **1990**, *53*, 129-139.
29. Love, W. G.; Havenaar, E. C.; Lowe, P. J.; Taylor, P. W. *SPIE Proc.* **1994**, *2078*, 381-388.
30. Zhou, C. *J. Photochem. Photobiol. B: Biol.* **1989**, *3*, 299-318.
31. Valduga, G.; Reddi, E.; Jori, G. *J. Inorg. Biochem.* **1987**, *29*, 59-65.
32. Jori, G. In *Photosensitizing Compounds: Their Chemistry, Biology, and Clinical Use;* Bock, G.; Harnett, S. , Eds.; Ciba Foundation Symposium, Wiley, Chichester, 1989, Vol. 146; pp 78-86.
33. Reddi, E.; Zhou, C.; Biolo, R.; Menegaldo, E.; Jori, G. *Br. J. Cancer* **1990**, *61*, 407-411.
34. Ginevra, F.; Biffanti, S.; Pagnan, A.; Biolo, R.; Reddi, E.; Jori, G. *Cancer Lett.* **1990**, *49*, 59-65.
35. Milanese, C.; Zhou, C.; Biolo, R.; Jori, G. *Br. J. Cancer* **1990;** *61*, 846-850.
36. Polo, L.; Reddi, E.; Garbo, G. M.; Morgan, A. R.: Jori, G. *Cancer Lett.* **1992**, *66*, 217-223.
37. Cuomo, V.; Jori, G.; Rihter, B.; Kenney, M. E.; Rodgers, M. A. J. *Br. J. Cancer* **1990**, *62*, 966-970.
38. Richter, A. M.; Waterfield, E.; Jain, A. K.; Canaan, A. J.; Allison, B. A.; Levy, J. G. *Photochem. Photobiol.* **1993**, *57*, 1000-1006.
39. Doiron, D. R.; Razum, N. J.; Trommer, R. M.; Syndu, R. S.; Morgan, A. M. *Proc. 5th Int. Photodynamic Assoc.* **1994**, 45.
40. Jori, G. *J. Photochem. Photobiol. B: Biol.* **1996**, *36*, 87-93.
41. Schieweck, K.; Capraro, H-G.; Isele, U.; van Hoogevest, P.; Ochsner, M.; Maurer, T.; Batt, E. *SPIE Proc.* **1994**, *2078*, 107-118.
42. Ochsner, M. *J. Photochem. Photobiol. B: Biol.* **1996**, *32*, 3-9.

43. Love, W. G.; van der Zanden, B. C. H.; Taylor, P. W. *Drug Dev. Ind. Pharm.* **1997,** *23,* 705-710.
44. Valduga, G.; Reddi, E.; Jori, G.; Cubeddi, R.; Taroni, P.; Valentini, G. *J. Photochem. Photobiol. B: Biol.* **1992,** *16,* 331-340.
45. Isele, U.; van Hoogevest, P.; Hilfiker, R.; Capraro, H-G.; Schieweck, K.; Leuenberger, H. *J. Pharm. Sci.* **1994,** *83,* 1608-1616.
46. Bartsch, W.; Sponer, G.; Dietmann, K.; Fuchs, G. *Artzneim. Forsch.* **1976,** *26,* 1581-1583.
47. Love, W. G.; Duk, S.; Biolo, R.; Jori, G.; Taylor, P. W. *Photochem. Photobiol.* **1996,** *63,* 656-661.
48. Isele, U.; Schieweck, K.; Kessler, R.; van Hoogevest, P.; Capraro, H-G. *J. Pharm. Sci.* **1995,** *84,* 166-173.
49. Gal, D.; MacDonald, P. C.; Porter, J. C.; Simpson, E. R. *Int. J. Cancer* **1981,** *28,* 315-319.
50. Norata, G.; Canti, G.; Ricci, L.; Nicolin, E.; Trezzi, E.; Catapano, A. L. *Cancer Lett.* **1984,** *25,* 203-208.
51. Rudling, M. J.; Reihnér, E.; Einarsson, K.; Ewerth, S.; Angelin, B. *Proc. Natl. Acad. Sci. USA* **1990,** *87,* 3469-3473.
52. Vitols, S.; Angelin, B.; Ericsson, S.; Gahrton, G.; Juliusson, G.; Masquelier, M.; Paul, C.; Peterson, C.; Rudling, M.; S"derberg-Reid, K.; Tidefelt, U. *Proc. Natl. Acad. Sci. USA* **1990,** *87,* 2598-2602.
53. Zhou, C.; Milanesi, C.; Jori, G. *Photochem. Photobiol.* **1988,** *48,* 487-492.
54. Rensen, P. C. N.; van Dijk, M. C. M.; Havenaar, E.; Bijsterbosch, M. K.; Kruijt, J. K.; van Berkel, T. J. C. *Nature Med.* **1995,** *1,* 221-225.
55. Versluis, A. J.; Rensen, P. C. N.; Kuipers, M. E.; Love, W. G.; Taylor, P. W. *J. Photochem. Photobiol. B:Biol.* **1994,** *23,* 141-148.
56. Polo. L.; Bianco, G. Reddi, E.; Jori, G. *Int. J. Biochem. Cell Biol.* **1995,** *27,* 1249-1255.
57. Rensen, P. C. N.; Love, W. G.; Taylor, P. W. *J. Photochem. Photobiol. B:Biol.* **1994,** *26,* 29-35.
58. Semple, S. C.; Chonn, A. *J. Liposome Res.* **1996,** *6,* 33-60.
59. Chapman, M. J. *J. Lipid Res.* **1980,** *21,* 789-853.
60. Love, W. G.; Rensen, P. C. N.; Isele, U.; Taylor, P. W. *SPIE Proc.* **1995,** *2371,* 187-193.

Chapter 12

Intracellular Delivery of Liposomal Contents Using pH- and Light-Activated Plasmenyl-Type Liposomes

Oleg V. Gerasimov, Nathan Wymer, Derek Miller,
Yuanjin Rui, and David H. Thompson[1]

Department of Chemistry, Purdue University, West Lafayette, IN 47907–1393

Applications of targeted liposomes presently suffer from inefficient contents release at the target site. We report plasma-stable plasmenylcholine and diplasmenylcholine liposome systems that release their contents upon exposure to low pH (≤ 6.5) or oxidative environments. Cell culture experiments also suggest that receptor-mediated uptake of folate-targeted diplasmenylcholine liposomes can efficiently deliver their contents to the cytoplasm of KB cells upon endosomal acidification.

Selective targeting and membrane translocation of drugs at specific anatomic sites are two of the most intractable problems in pharmaceutics. Liposomal drugs have attracted a great deal of attention as drug delivery vehicles due to their favorable biocompatibility, ease of production, high drug:lipid ratios, blood clearance characteristics, and targetability (via active [1,2] or passive [3] methods), however, the membrane translocation issue (i.e. penetration of the 4 nm hydrophobic barrier of the target cell membrane) has not yet been solved. The inefficient intracellular delivery of peptide, antisense oligonucleotide, and gene constructs to target cells is due, in part, to the limited membrane exchange that occurs with materials of this size and hydrophilicity. Endosomal delivery pathways can result in efficient intracellular uptake of macromolecules, however, materials internalized in this manner often remain localized within the endosome and are ultimately degraded within the lysosome before the contents can escape unless specific release mechanisms have been incorporated. There remains a need for intelligent, triggerable materials for delivery of labile hydrophilic macromolecules that respond to changes in the local environment of the drug carrier such that it releases its payload in the vicinity of its target in a controlled manner.

There are at present a limited number of effective, biocompatible methods for triggering contents release and membrane fusion from plasma-stable liposomes [4]. The most widely studied triggering approaches utilize either low pH, thermal, or photochemical triggering. The photoinduced and pH-induced release properties of liposomes formed from plasmenylcholine and diplasmenylcholine lipids is the subject of this article. Opportunities for the application of these liposomes in intelligent drug delivery systems is also briefly discussed.

Plasmenylcholine and Diplasmenylcholine Lipids

Phospholipids containing O-1′-alkenyl linkages to the glycerol backbone belong to the

[1]Corresponding Author.

class of lipids commonly known as plasmalogens. Plasmenylcholines are derivatives of this class that contain a single O-1'-Z-alkenyl chain at the glycerol sn-1 position and a 3-phosphocholine headgroup (Figure 1). Plasmalogen lipids are widely distributed in mammalian tissues, comprising up to 65% of the phospholipid fraction in the electrically-active cells [5]. Although their functional role in biomembranes is not entirely clear, they are known to possess an unusually high proportion of arachidonate [6], suggesting that they may be involved in signal transduction pathways. Diplasmenylcholines, in contrast, contain O-1'-Z-alkenyl chains at both the sn-1 and sn-2 positions and have been found in only the acrosomal membranes of rabbit sperm cells [7]. Previous efforts from this group have described the first stereospecific total syntheses of plasmenylcholines [8] and diplasmenylcholines [9] that are sufficiently general to provide access to a wide variety of 1'- and 2'-modified derivatives. The electron-rich vinyl ether linkage of these lipids is susceptible to either oxidation or acid-catalyzed hydrolysis, producing single chain surfactants (i.e. lysophosphatidylcholine and fatty aldehyde in the case of plasmenylcholine [10] or two equivalents of fatty aldehyde in the case of diplasmenylcholine [11]; Figure 2). Either degradative reaction may induce localized micellar-like defect structures within a host bilayer membrane (Figure 3) as suggested by the lipid packing model [12]. It is this chemically-induced phase transition feature that we exploit within the context of environmentally-responsive liposomal drug delivery.

Photooxidative Triggering of Plasmenyl-Type Liposomes

Light-activation is a promising approach for controlling the site and rate of drug release from parenteral drug carriers, particularly when wavelengths in the 700-1100 nm range are used where tissue penetration depths approach 1 cm [4]. Photochemical generation of oxidants such as singlet oxygen ($^1O_2$) can be achieved by near-infrared illumination of a suitable sensitizer in the presence of O_2. Singlet oxygen, in turn, can then be used to oxidize the O-1'-Z-alkenyl bond of plasmenyl-type lipids via a putative 1,2-dioxetane intermediate. Subsequent cleavage of this intermediate results in the generation of bilayer-disrupting single chain species that facilitate escape of the internalized material from the liposome carrier. Bacteriochlorophyll a (Bchl$a$), a commercially-available, membrane-soluble dye that efficiently generates $^1O_2$ in aerobic solutions upon excitation of its 780 nm absorption maximum, was chosen as the sensitizer for exploring the photooxidative triggering properties of plasmenylcholine and diplasmenylcholine liposomes at neutral pH near physiological temperature.

Bacteriochlorophyll a-Sensitized Release of a Model Hydrophilic Drug from Plasmenylcholine Liposomes.
The photorelease characteristics of Bchl$a$:2-palmitoyl-sn-plasmenylcholine (PlsPamCho) liposomes were monitored using calcein (a water-soluble fluorescein derivative) as a model hydrophilic drug [13]. The observed release rates (Figure 4) were found to be dependent upon (i) light intensity, (ii) oxygen concentration, (iii) the presence of plasmenylcholine lipid within the liposomal membrane, and (iv) the absence of singlet oxygen quenchers. Morphological changes occurring in the photolyzed samples further suggest that membrane-membrane fusion proceeds on a similar timescale with contents leakage [13]. These data suggest that site-specific drug release may be possible using appropriately designed, phototriggerable plasmenylcholine liposomes, particularly if they have been initially targeted using either active or passive localization methods.

Bacteriochlorophyll a-Sensitized Release of Ca^{2+} from Diplasmenylcholine Liposomes.
The photooxidative triggering pathway was also explored in dipalmitoyl-based diplasmenylcholine (DPPlsC) liposomes containing Ca^{2+} in the intraliposomal compartment and Bchl$a$ partitioned into the bilayer. Figure 5

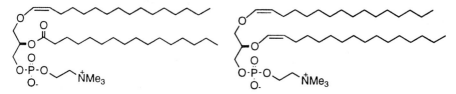

Figure 1. Structures of plasmenylcholine (left) and diplasmenylcholine (right).

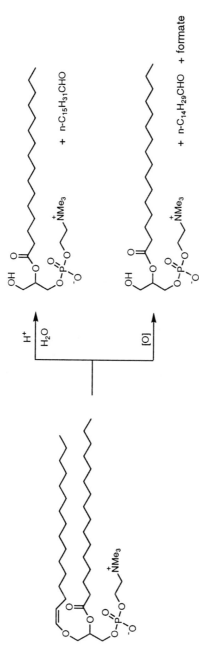

Figure 2. Hydrolysis (top) and photooxidation (bottom) reactions of plasmenylcholine.

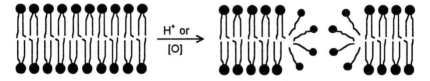

Figure 3. Single chain surfactant-induced lamellar-micellar phase transition.

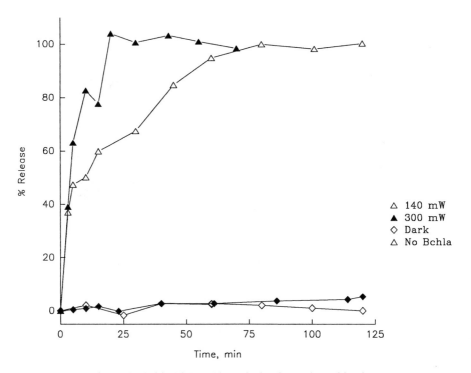

Figure 4. Bchl*a*:PlsPamCho calcein photorelease kinetics.

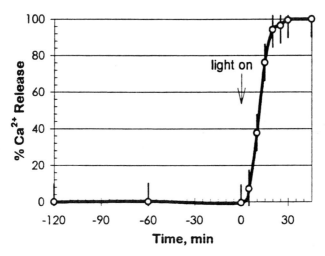

Figure 5. Bchl*a*:DPPlsC Ca²⁺ photorelease kinetics.

shows that calcium ions are retained by these liposome formulations in the dark. Photolysis of these samples, however, causes rapid calcium ion efflux. When similar experiments are performed in the presence of EDTA, the free calcium ion concentrations remain low until the photoreleased Ca^{2+} concentration exceeds the external EDTA concentration. These results suggest that calcium ions could be used in a "cascade-type" liposome triggering reaction (Figure 6). Appropriately designed Bchla:DPPlsC liposomes containing Ca^{2+} would be used as a catalytic subpopulation of light-activatable liposomes that would trigger a larger population of drug-loaded conventional liposomes using a Ca^{2+}-dependent catalytic reaction. We have recently demonstrated this principle by photoreleasing Ca^{2+} from Bchla:DPPlsC liposomes to regulate the activity of PLA_2. PLA_2 hydrolysis of 1,2-dipalmitoyl-*sn*-phosphocholine (DPPC) contained within the second population of liposomes leads to DPPC liposome contents leakage via a cascade reaction mechanism [14]. Since DPPlsC is not hydrolyzed by PLA_2, cascade triggering may be particularly valuable for promoting endosomal release and cytoplasmic delivery of bioactive macromolecules (e.g. plasmids) from multifunctional liposomes that contain Ca^{2+}, PLA_2, and the bioactive agent inside. Although the concept is an attractive "next-generation" triggering technique that may not require significant reformulation of liposomal drugs that have already received FDA approval, safety concerns associated with the use of PLA_2 in a cascade triggering scheme may limit its suitability for intravascular applications as presently formulated. This potential problem might be obviated by [i] deposition via other parenteral routes (i.e. intramuscular or intraperitoneal), [ii] immobilization of PLA_2 within a liposome-containing hydrogel, [iii] inclusion of PLA_2-triggerable cascade formulations within implantable hollow-fiber reactors whose MW cutoff would prevent escape of the enzyme, and [iv] encapsulation of a specific PLA_2 inhibitor along with the drug-loaded subpopulation of DPPC liposomes such that their hydrolysis leads to eventual enzyme poisoning.

Acid-Catalyzed Triggering of Plasmenyl-Type Liposomes

Of the triggering modalities currently under investigation in liposome- [4] and polymer-based [15] drug delivery systems, low pH environments are the most biochemically accessible due to their prevalence in endosomal compartments and ischemic tissues. One pathway for intracellular drug delivery, shown in Figure 7, involves targeting of the drug carrier via a ligand that is internalized by receptor-mediated endocytosis. Once inside the endosome, acidification of this compartment proceeds to lower the internal pH to values approaching 4.5-5.0 in some cell types. Drug carriers appropriately designed to undergo membrane fusion and contents release at low pH, therefore, may be capable of delivering their cargo directly to the cytoplasm of the target cell if they are imported via this endocytosis process.

pH-Dependent Hydrolysis Kinetics of Diplasmenylcholine and Its Relationship to Liposomal Release Rates. We have undertaken an extensive study of triggered calcein release from PlsPamCho and DPPlsC liposomes in acidic buffers to serve as a simplified model system for the endosomal release process [10,11]. The rates of calcein release are strongly dependent on solution pH, temperature, and membrane composition. PlsPamCho liposomes release calcein more slowly in the presence of dihydrocholesterol (DHC; $t_{50\% \text{ release}}$ = 3 min and 65 min for pure PPlsC and 6:4 PPlsC:DHC liposomes, respectively, at pH 2.5) [10] and with rates that are pH-dependent (Table I).

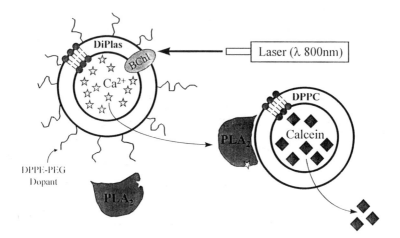

Figure 6. Cacasde photorelease scheme.

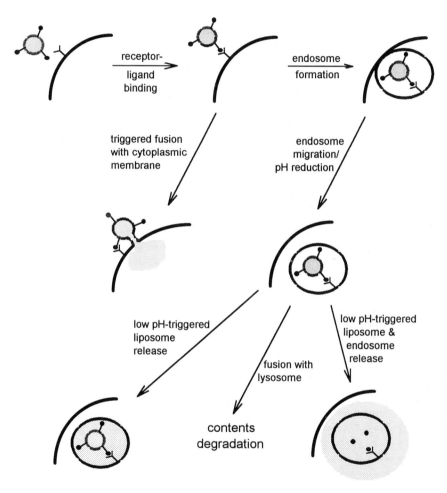

Figure 7. Model for endosomal targeting, uptake and release.

Table I.   pH-Dependence on Calcein Release from PlsPamCho
Liposomes.

pH	$t_{50\% \text{ release}}$
2.3	4 minutes
3.2	9 minutes
4.3	70 minutes
5.3	9 hours
6.3	~4 days
7.4	> 5 weeks

Parallel studies of lipid hydrolysis and calcein release indicate that a greater extent of hydrolysis is required to effect 50% contents release when DHC is present (< 4% and ~28% hydrolysis for pure PPlsC and 6:4 PPlsC:DHC liposomes, respectively) [10]. Sphingomyelin (SM) or 1,2-distearoyl-sn-phosphatidylcholine (DSPC) liposomes can also be rendered pH-sensitive by including DPPlsC as a minor component (20-40 mol%) in the formulation, although, the calcein leakage kinetics are somewhat slower in these systems ($t_{50\% \text{ release}} \cong 120$ min for 8:2 DSPC:DPPlsC and 6:4 SM:DPPlsC at pH 4.5; Figure 8).   DPPlsC and PlsPamCho liposomes are qualitatively similar in their leakage characteristics (i.e. DHC suppresses, while lysolipid enhances, calcein effusion rates) [10].   One of the most striking differences between the PPlsC and DPPlsC hydrolysis behavior, however, is the sudden onset of calcein leakage with small changes in the extent of lipid hydrolysis for the case of DPPlsC, regardless of the mol% DHC present (Figure 9).   This threshold response suggests that a phase transition occurs during DPPlsC hydrolysis, such that calcein leakage occurs rapidly from this new phase once a critical amount of hydrolysis products have accumulated within the bilayer.

Folate-Targeting to KB Cells & pH-Induced Endosomal Release. Targeting of egg phosphatidylcholine (EPC) liposomes to folate receptor-bearing cells has recently been reported [16].   Folate-deficient culture media and approximately ten folate ligands covalently attached to the liposome via a MW 3350 polyethylene glycol (PEG)-distearoylphosphatidylethanolamine conjugate (folate-PEG-DSPE) are required for efficient cell targeting.   Folate-targeted liposomes are processed via a typical endosomal uptake mechanism, since folate-PEG-DSPE:EPC liposomes containing encapsulated calcein showed punctate cellular fluorescence, indicative of   cellular uptake by (but not release from) the endosomes, within 4 h at 37°C [17].   This evidence suggests that endosomally-targeted materials do not escape from the endosome in the absence of a triggerable membrane fusion pathway.   A pH-sensitive, fusogenic liposome construct, therefore, would be ideal for endosomally-targeted materials.

Folate-targeted, pH-sensitive DPPlsC liposomes were tested for their ability to release propidium iodide (PI) into the cytoplasm of KB cells upon endosomal uptake and acidification [11].   Propidium iodide release kinetics, determined by fluorometric ratio assay using 540 nm excitation while monitoring emission at 615 nm [18], reveal that ≥ 80% of the dye escaped both the liposomal and endosomal compartments within 8 hours when 0-10 mol% DHC was incorporated in the DPPlsC liposome membrane (Table II).

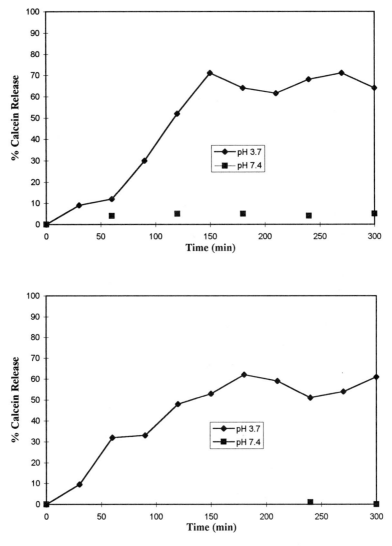

Figure 8. Calcein release kinetics from DPPlsC liposomes containing 60 mol% sphingomyelin (top) and 80 mol% distearoylphosphatidylcholine (bottom) at pH 3.7 and 7.4.

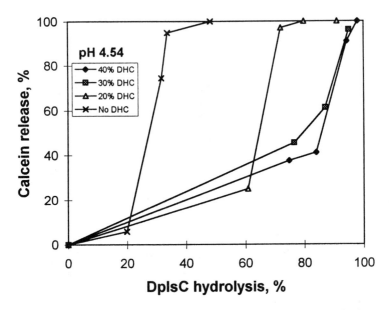

Figure 9. DPPlsC calcein release kinetics as a function of lipid hydrolysis at pH 4.5.

Table II. Release Efficiency of Propidium Iodide into the Cytoplasm of KB Cells Using Folate-Targeted Liposomes (0.5 mol% Folate-PEG-DSPE).

Liposome Composition	% PI Release after 8 h
EPC (control)	< 3%
DPPlsC	83%
9:1 DPPlsC:DHC	81%
8:2 DPPlsC:DHC	36%

Control experiments indicated that (*i*) DPPlsC liposomes are stable for more than 24 hours in the folate-deficient minimal Eagle's medium used for the assay and (*ii*) less than 5% PI release was observed from DPPlsC:DHC:folate:PI liposomes incubated with KB cell suspensions in the presence of the endosomal acidification inhibitors, chloroquine (50 μM) or monensin (25 μM) for 24 hours. These results clearly indicate that pH triggering with DPPlsC liposomes is a practical and efficient method for intracellular delivery of biologically active materials. The mechanistic basis for the speed and efficiency of this system is currently under investigation.

Conclusions

Plasmenyl-type liposomes are an interesting new class of liposomes that can be formulated in many different ways to confer biocompatible pH- or light-triggerability in a plasma-stable liposomal drug carrier. The vinyl ether cleavage mechanisms on which these liposomes are based are also under investigation for their ability to induce membrane-membrane fusion in poly(ethylene glycol)-modified "spontaneous" liposome formulations [*19*].

Acknowledgements. The authors would like to acknowledge the support of the Whitaker and Purdue Research Foundations, INEX Pharmaceuticals, and the National Institute of Health (Shannon Award).

Literature Cited

1. Leserman, L.; Suzuki, H.; Machy, P. In *Liposome Technology*; Vol. III, 2nd ed., Gregoriadis, G., Ed.; CRC Press: Boca Raton, FL, 1993; pp. 139-152.
2. Mori, A.; Huang, L. In *Liposome Technology*; Vol. III, 2nd ed., Gregoriadis, G., Ed.; CRC Press: Boca Raton, FL, 1993; pp. 153-162.
3. Stealth Liposomes, Lasic, D. D.; Martin, F. J., Eds., CRC Press: Boca Raton, FL, 1995.
4. Gerasimov, O. V.; Rui, Y.; Thompson, D. H., In *Vesicles*; Rosoff, M., Ed.; Marcel Dekker: New York, NY, 1996; pp. 679-746.
5. Paltauf, F. *Chem. Phys. Lipids* **1994** *74*, 101.
6. Snyder, F.; Lee, T.-C.; Blank, M. L. *Prog. Lipid Res.* **1992** *31*, 65.
7. Touchstone, J. C.; Alvarez, J. G.; Levin, S. S.; Storey, B. T. *Lipids* **1985** *20*, 869.
8. Rui, Y.; Thompson, D. H. *Chem. Eur. J.* **1996** *2*, 1505.
9. Rui, Y.; Thompson, D. H. *J. Org. Chem.* **1994** *59*, 5758.
10. Gerasimov, O. V.; Schwan, A.; Thompson, D. H. *Biochim. Biophys. Acta* **1997** *1324*, 200.
11. Rui, Y.; Wang, S.; Low, P. S.; Thompson, D. H., submitted.
12. Israelachvili, J. N. *Intermolecular & Surface Forces*; Wiley: New York, NY, 1985; pp. 256.

178

13. Thompson, D. H.; Gerasimov, O. V.; Wheeler, J. J.; Rui, Y.; Anderson, V. C. *Biochim. Biophys. Acta* **1996** *1279*, 25.
14. Wymer, N.; Gerasimov, O. V.; Thompson, D. H., submitted.
15. Hoffman, A. *Macromol. Symp.* **1995** *98*, 645.
16. Lee, R. J.; Low, P. S. *J. Biol. Chem.* **1994** *269*, 3198.
17. Lee, R. J.; Ph. D. Thesis, Purdue University, **1994**, pp. 8-28.
18. Vogel, K.; Wang, S.; Lee, R. J.; Chmielewski, J.; Low, P. S. *J. Am. Chem. Soc.* **1996** *118*, 1581.
19. Szleifer, I.; Gerasimov, O. V.; Thompson, D. H. *Proc. Natl. Acad. Sci. USA*, in press.

Chapter 13

Conditionally Stable Liposomes for Gene Delivery

S. C. Davis and F. C. Szoka, Jr.[1]

Department of Biopharmaceutical Sciences and Pharmaceutical Chemistry, School of Pharmacy, University of California, San Francisco, CA 94143–0446

A novel class of triggered release liposomes is described. A series of phosphate monoester lipids have been synthesized which form stable liposomes with dioleoylphosphatidylethanolamine but which collapse on exposure to alkaline phosphatase. Transfection of kidney fibroblast cells in vitro highlights the potential utility of this system for gene therapy.

One of the more promising methods for the delivery of drugs and DNA to the cytoplasm of cells is the use of triggered release liposomes (*1*). These liposomes are stable under normal physiological conditions but release their contents upon receipt of a defined stimulus. This allows the controlled delivery of the liposome contents to the site of the stimulus, either intracellularly or at defined sites in the body.

Triggered release liposomes rely on the ability of lipids to assume different phases in aqueous media and to interconvert between them. The two phases most commonly used are the lamellar phase and the inverted hexagonal phase. The lamellar phase of liposomes and cell membranes can encapsulate exogenous material in vesicles whereas the inverted hexagonal phase is incapable of doing so. Thus collapse of a lamellar phase liposome to the inverted micelle array of the inverted hexagonal phase or any other phase leads to release of the initial liposomes' contents (*2*).

Phosphatidylethanolamine lipids such as dioleoyl phosphatidylethanolamine (DOPE) naturally form the inverted hexagonal phase under physiological conditions but by addition of a charged colipid the phosphatidylethanolamine can be forced into the lamellar phase (*3*). Removal of the charged colipid or merely the charged portion allows the phosphatidylethanolamine to collapse into the hexagonal phase with concomitant release of the liposome contents and potentially membrane fusion (*1, 2*).

[1]Corresponding Author.

Several different strategies have been used for the controlled removal of charge from the surface of a phosphatidylethanolamine/colipid bilayer including pH change (e.g. cholesterol hemisuccinate, CHEMS (4)), enzyme catalysis (e.g. phospholipase C or acetylcholine esterase (5)), and light (cleavage of plasmalogen vinyl ethers (6)).

Freeze fracture electron microscopy can be used to follow the phase transitions involved in the collapse. The system captured in the following three pictures (Figure 1) is the pH sensitive liposome, CHEMS/TPE (Transesterified Egg phosphatidylethanolamine) (7). Panel A shows the liposomes at pH 7.4 and 25°C, under these conditions CHEMS is charged and liposomes are observed. In panel B the system is shown at pH 4.5 and 4°C, where the carboxylic acid of CHEMS is protonated. The liposomes are aggregating and becoming larger but the low temperature prevents the phase transition to the hexagonal phase. Panel C shows the effect of allowing the temperature to rise to 25°C, the liposomes have completely collapsed to the inverted hexagonal phase and no discrete liposomes are observed.

Design of a Novel Triggered Release Liposome

We hypothesized that the phosphate monoesters of lipids (phosphatidic acids), which are doubly negatively charged species, will stabilize phosphatidylethanolamine bilayers. Phosphatase enzymes which catalyze the hydrolysis of phosphate monoesters could therefore remove the phosphate and corresponding charge from the bilayer surface allowing collapse to the hexagonal phase and release of the liposome contents. Two examples of this class of enzymes are alkaline phosphatase (8), which is a widely specific, highly active enzyme found in serum and on the cell membrane, and acid phosphatase which is found intracellularly in endosomes and lysosomes (9).

As the structural requirements for the phosphatidic acid colipid needed for phosphatase mediated triggered release liposomes were unknown, a flexible synthesis was designed. The synthesis allows us to modulate the distance of the phosphate group from the bilayer, the hydrophobicity of the lipid and the number of heteroatoms in the linking chain. Cholesterol was chosen as the hydrophobic core of the lipid for synthetic ease, our experience with CHEMS (4) and due to its ability to stabilize bilayers to the action of serum (10). Diacyl glycerol lipids are also being investigated. The key step in the synthesis is the $S_N 1$ type reaction of cholesterol tosylate with a diol (11) to yield an alcohol of the structure shown in Figure 2. The reaction is relatively insensitive to the structure of the diol and so a large number of lipids can be synthesized. The alcohol can then be phosphorylated under standard conditions (12).

Initial screen of liposome formation The precursor alcohols were tested for the ability to form bilayers with a colipid. The premixed lipids were dried to a film under vacuum and rehydrated in 10 mM HEPES pH 7.4, 5% glucose with vortexing. The suspension was then diluted with 10 mM HEPES pH 7.4, 150 mM NaCl on a microscope slide and the structures inspected visually. The first colipid studied was egg phosphatidylcholine (Egg PC) which readily forms liposomes under most conditions. This quick screen was then performed using DOPE in place of Egg PC at pH 9.0, where DOPE is partially charged and will support a bilayer. Those lipids which formed liposomes under these conditions were then diluted into pH 7.4 buffer and the outcome observed. The results of this screen are shown in Figure 2.

Panel A

Figure 1. Freeze fracture electron micrographs of CHEMS/TPE (1:3) liposomes under the following conditions. Panel A: pH 7.5 at 25°C. Panel B: pH 4.5 at 4°C. Panel C: pH 4.5 at 25°C. Bar is 0.2 μm.

Continued on next page.

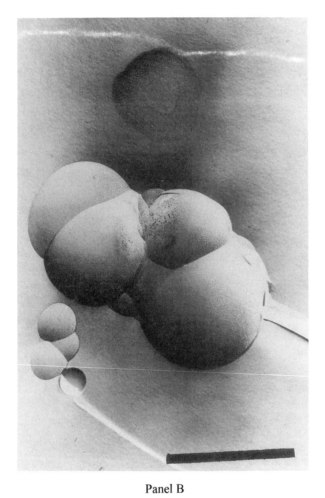

Panel B

Figure 1. *Continued.*

Panel C

Figure 1. *Continued.*

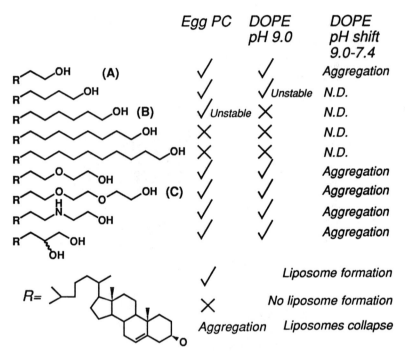

Figure 2. The effect of cholesterol alcohol structure upon liposome formation. The lipids were present in a 4:5 ratio of cholesterol derivative to colipid.

A selection of the alcohols we have synthesized are shown and certain trends can be observed. Egg PC supports liposome formation with all but the most hydrophobic alcohols whereas DOPE is more sensitive to sterol structure, even at pH 9. Only the heteroatom containing sidechains and the shortest alcohols support a bilayer structure. Protonation of the DOPE at pH 7.4 causes collapse of all the systems studied, in an analogous fashion to that expected for removal of phosphate from the final system.

From this data we decided to initially test three phosphates: these were the phosphates synthesized from the cholesterol-C2 alcohol (**A**), the cholesterol-C6 alcohol (**B**), and the cholesterol-triethylene glycol (**C**). The three alcohols were chosen due to their differing behavior with DOPE and structural diversity. The phosphates derived from these alcohols will position the charges at varying distances from the membrane interface and have different levels of hydrophobicity.

Action of Alkaline Phosphatase on Liposomes. In contrast to the precursor alcohols the phosphorylated lipids smoothly formed liposomes with DOPE at pH 7.4, which were sonicated and extruded to give small unilamellar liposomes. Carboxyfluorescein was used as a self quenched fluorescent marker in these liposomes and dye leakage was measured (Figure 3). Suspension of the liposomes in buffer (HBS) caused little dye leakage for all three lipids (only one is shown for clarity), however addition of alkaline phosphatase to the liposomes caused an enhanced leakage of liposome contents.

Further experiments were done with the most sensitive preparation, the cholesterol C2 phosphate/DOPE liposome, in order to show that the enzyme was causing the leakage via the proposed catalytic dephosphorylation of the liposome. The rate of leakage was shown to be dependent upon the amount of enzyme activity present in the reaction and that heat inactivated enzyme had no effect upon the liposome. Incubations performed with a small amount of enzyme showed a significant lag phase before leakage was observed (data not shown). This is consistent with a stepwise removal of charge from the liposome surface until the liposomes spontaneously collapse rather than a non specific bilayer-protein interaction..

Alkaline phosphatase catalyses the hydrolysis of phosphate monoesters to inorganic phosphate and an alcohol. Therefore we should detect both cholesterol C2 alcohol (**A**) and inorganic phosphate after exposure of the liposomes to the enzyme. Bligh-Dyer extraction of the lipids from the aqueous phase after a leakage experiment enabled us to detect the alcohol A in the organic phase using a t.l.c. assay. Inorganic phosphate remained in the aqueous phase where it was measured after incubation in the presence of enzyme.

These experiments show that alkaline phosphatase catalyzes the triggered release of the liposomes contents as proposed in the initial hypothesis.

In Vitro Transfection

Encapsulation of DNA In order to test the utility of this novel liposome system, transfection experiments were performed in vitro. Encapsulation of DNA in such negatively charged liposomes was potentially a problem but was readily achieved utilizing a modification of the REV methodology (*13*). The premixed lipids in chloroform were dried down to a lipid film. This was redissolved in a mixture of diethyl ether and 1,1,2-trichloro-2,2,1-trifluoroethane (57:43, 3 ml), which has a density of 1.06, and buffer (0.5 ml) was added. This was vigorously sonicated to form an emulsion and then the DNA (0.5 ml) was added followed by further brief sonication. The emulsion was then dried down to the gel phase which

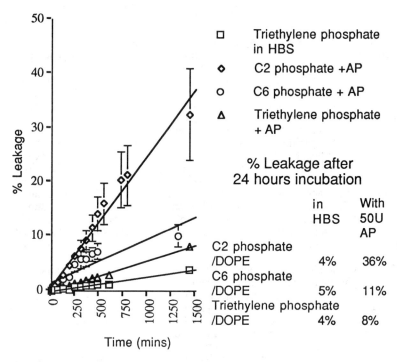

Figure 3. The effect of alkaline phosphatase upon liposome leakage in HEPES buffered saline (HBS).

was broken under reduced pressure to yield liposomes. These liposomes were purified on a discontinuous Ficoll gradient to remove excess DNA.

Analysis of the liposomes showed a bimodal size distribution with peaks at 300 nm and at 80 nm. 25 % of the initial DNA was associated with the liposomes using this protocol and DNase protection assays showed that the DNA was encapsulated inside the liposomes or otherwise protected from degradation.

Transfection of monkey kidney fibroblast cells CV-1 cells were plated in a 96 well plate at 2×10^4 cells per well in 100 µl of growth medium consisting of DME-H21 with 10% fetal calf serum and antibiotics. Cells were then incubated for 19 hours at 37°C before the medium was removed and replaced with serum free media or media containing 10% serum. Liposomes encapsulating a luciferase expression vector (pLC 0626, GeneMedicine Inc.) were then added to this media (0.5 µg DNA) and incubated for 5 hours at 37°C. The medium was then removed and replaced with growth media and the cells incubated for a further 48 hours. Luciferase activity and protein concentration were then measured (Figure 4). Little toxicity was observed with this system in any of the eight occasions the experiment was repeated. The results of a representative experiment are shown. Naked DNA was incapable of transfecting this cell line, whereas a heat degraded polyamidoamine cationic dendrimer (14) was a very efficient transfection agent.

The transfection efficiency of the cholesterol C2 phosphate/ DOPE liposomes is comparable to that seen with other triggered release systems such as CHEMS/DOPE under these conditions (4). The mechanism of transfection is not yet clear and further experiments to investigate this matter are in progress. The cationic dendrimer is significantly more efficient than any triggered release system in the absence of serum in vitro, perhaps due to increased absorption at the cell surface (4). The cholesterol C2 phosphate/DOPE liposomes appear to be mildly sensitive to serum in this assay (10 fold reduction in transfection).

Conclusion

We have proposed and synthesized a novel class of triggered release liposomes . These conditionally stable liposomes collapse in the presence of alkaline phosphatase leading to the release of the liposome contents. DNA encapsulated in these liposomes is capable of transfecting CV-1 cells in vitro at levels competitive with other triggered release systems. Further work to characterize and improve this triggered release liposome system for the delivery of DNA in vitro and in vivo is currently in progress.

Acknowledgments This work was supported by NIH grants GM 26691, DK 46052, DK 47766. Dr. Szoka has a financial interest in and serves as a consultant to GeneMedicine, Inc. a biotechnology company developing gene medicines. We are grateful to Dr. Ming-Zong Lai for the freeze fracture electron micrographs.

References

1. Gerasimov O. V.; Rui Y.; Thompson D. H. In *Vesicles*; Rosoff M., Ed.; Marcel Dekker: New York, NY **1996**; 679-740.
2. Chu C-J.; Szoka F.C. *J. Liposome Res.* **1994**, *4*, 361-395.
3. *CRC Handbook of Lipid Bilayers*, Marsh D., Ed.; CRC Press, Boca Raton, FL., **1990**; 128-130, 266-267.
4. Legendre J-Y.; Szoka F. C. *Pharm. Res.* **1992**, *9*, 1235-1242.

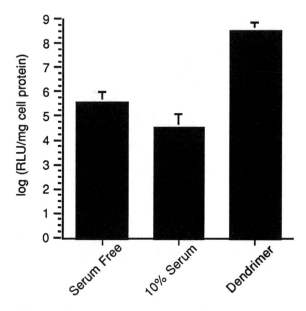

Figure 4. Transfection of CV-1 cells with a luciferase expression plasmid. Serum free and 10% serum experiments refer to the cholesterol-C2 phosphate/DOPE liposomes.

5. Menger F. M.; Johnston D. E. *J. Am. Chem. Soc.* **1991**, *113*, 5467-5468.
6. Anderson V. C.; Thompson D. H. *Biochim. Biophys. Acta* **1992**, *1109*, 33-42.
7. Lai M-Z.; Vail W. J.; Szoka F. C. *Biochemistry* **1985**, *24*, 1654-1661.
8. Millàn J. L.; Fishman W. H. *Crit. Rev. Clin. Lab. Sci.* **1995**, *32*, 1-39.
9. deDuve C.; Pressman B. C.; Gianetto R.; Wattiaux R.; Appelmans F. *Biochem. J.* **1955**, *60*, 605-617.
10. Liu D.; Huang L. *Biochim. Biophys. Acta* **1989**, *981*, 254-260.
11. Patel K. R.; Li M. P.; Schuh J. R.; Baldeschweiler J. D. *Biochim. Biophys. Acta* **1985**, *814*, 256-264
12. Danilov L. L.; Maltsev S. D.; Shibaev V. N. *Bioorganicheskaya Khimiya* **1990**, *16*, 1002-1003.
13. Szoka F.; Papahadjopoulos D. *Proc. Nat. Acad. Sci. USA* **1978**, *75*, 4194-4198.
14. Tang M. X.; Redemann C. T.; Szoka F. C. *Bioconjugate Chem.* **1996**, *7*, 703-714.

Chapter 14

Fundamental Studies of DNA Adsorption and Hybridization on Solid Surfaces

Lynn Anne Sanguedolce[1], Vincent Chan[1], Steven E. McKenzie[1,2], Saul Surrey[2], Paolo Fortina[3], and David J. Graves[1]

[1]Department of Chemical Engineering, University of Pennsylvania, 311A Towne Building, 220 South 33 Street, Philadelphia, PA 19104–6393
[2]Departments of Pediatrics and Research, Jefferson Medical College, Philadelphia, PA and the duPont Hospital for Children, Wilmington, DE
[3]Department of Pediatrics, The University of Pennsylvania and Children's Hospital of Philadelphia, Philadelphia, PA 19104

Arrays of DNA or oligonucleotide sequences immobilized on a solid surface are becoming important tools for identifying genetic defects, pathogens, forensic samples, gene expression profiles, etc. by hybridization to solution-phase complements. However, little fundamental work has been done on such areas as solution and surface diffusion, steric hindrance due to the solid support, immobilized probe density, spacer arm effects, and other parameters which can affect hybridization. Our current results include data on adsorption equilibria and desorption rates of oligonucleotide on various types of silanized surfaces, surface diffusion constants, and preliminary data on spacers and the effect of sequences on either side of the hybridization region. Surface concentrations which cause fluorescence quenching are surprisingly easy to attain, indicating close packing and revealing a potential problem which may exist in certain situations.

Arrays of oligonucleotides, PCR products and cDNA of different sequences are beginning to find important applications in bioscience. These arrays contain single-stranded nucleotide chains (probes) typically 10 to 20 nucleotides in length or longer, that are, in their most useful form, arrayed and attached covalently on a surface such as glass or silicon. The arrays are then contacted with mixture of targets, single-stranded DNA or RNA molecules from a sample to be studied that have been labeled (usually with a fluorescent tag). Under proper conditions, complementary sequences will hybridize. From a knowledge of the sequence of each probe and its location in the array, and from measurements of which probes fluoresce, one can determine the sequences present in the unknown sample. This provides a powerful tool for studying cells and their metabolism

Although conventional technology has been used to place 96 or 384 such probes in a cm-scale pattern on nylon membranes, some of the more interesting and potentially useful applications involve thousands to hundreds of thousands of different spots in a checkerboard pattern only a few mm square. Among the analyses that can be performed with such a system are the rapid identification of pathogens, monitoring of gene expression profiles(1,2), screening for gene mutations and polymorphisms (3-7), mapping of genomic and cDNA clones(8), identification of a particular cell for forensic purposes, studies of evolution, changes in gene copy number, and many others.

This technology has numerous advantages. The parallel nature and large number of tests that can be performed simultaneously reduce labor dramatically and make possible analyses which would otherwise be prohibitive in cost. In addition, the small size scale reduces cost of expensive reagents and waste generated. Although this technology is developing rapidly in a number of industrial and university research laboratories, some of the basic questions questions concerning how the process works and what its limitations are have not been addressed. In this report we examine fundamental issues of adsorption, diffusion, and hybridization of oligodeoxy–nucleotides.

Materials and Methods

Oligonucleotide Probes and Targets Oligonucleotides were synthesized using standard phosphoramidite chemistry and purified by either HPLC or PAGE (Midland Certified Reagent Company, Midland, TX), unless noted differently. Oligonucleotides were synthesized with a 3'-amino group for attachment to glass, (Glen Research, Sterling, VA or Clontech, Palo Alto, CA) and are referred to as probes. In one set of experiments, probes having both a 3' -amino and a 5'-fluorescein group were used (Glen Research or Peninsula Laboratories, Inc., Belmont, CA). Oligonucleotide probes from the ß-globin gene have been described previously (9) and are listed in Table I. Solution-phase target molecules were 5' -labeled with fluorescein for hybridization studies.

Cleaning and silanizing glass supports Microscope cover glasses (Corning No. 1 1/2, 25 square mm, Fisher, Pittsburgh, PA) were acid-cleaned in a class 100 clean room with a 1:2 (v/v) solution of 30% (w/w) hydrogen peroxide and 96% (w/w) sulfuric acid (Clean Room Electronics Chemicals, EMSCO, Philadelphia, PA) for 10 min at 120°C, rinsed in deionized water for 5 min and dried on a hot plate for 5 min at 110°C (10). The cleaned glass was silanized in 1% (v/v) 3-aminopropyl triethoxysilane (APTES) (Sigma, St. Louis, MO) in 95% (v/v) ethanol (Pharmco Products Inc., Brookfield, CT) (11-13), which leaves reactive amine groups attached to the surface. The solution was neutralized with acetic acid to achieve a high silane loading on the surface (14). The glass was then treated for 20 min at room temperature, rinsed in 100% ethanol and dried in vacuo at 110°C for 20 min. These surfaces either were then used for adsorption and diffusion experiments, or were further modified to covalently couple oligonucleotides to glass for surface density and hybridization studies.

Conversion of the amino-glass surface Two different methods, each using a homo-bifunctional crosslinking agent, were tested to attach probes to the modified glass surface. One involved use of glutaraldehyde and the other 1,4-phenylene diisothiocyanate (PDC). In each case, the amino-functional glass surface was first converted to a modified reactive surface that could couple the oligonucleotides. For gluteraldehyde modification, APTES-modified glass was treated with 12.5% (w/v) glutaraldehyde (Sigma) in coupling buffer (0.1 M Na_2HPO_4, 0.15 M NaCl, pH 7.0) with 0.6 g of sodium cyanoborohydride (Sigma) per 100 ml of coupling buffer (15). The glass was immersed in the glutaraldehyde solution for 4 h at room temperature, followed by rinsing in coupling buffer, deionized water and drying in vacuo for 20 min.

Table I

Function of Sequence	Sequence
Normal ß6	5' CCT GAG GAG AAG TCT 3'
Mutant ß6	5' CCT G<u>T</u>G GAG AAG TCT 3'
Normal ß39	5' TTG GAC CCA GAG GTT 3'
Mutant ß39	5' TTG GAC C<u>T</u>A GAG GTT 3'
Normal ß6 reverse complement	5' AGA CTT CTC CTC AGG 3'
Mutant ß6 reverse complement	5' AGA CTT CTC C<u>A</u>C AGG 3'
Normal ß39 reverse complement	5' AAC CTC TGG GTC CAA 3'
Mutant ß39 reverse complement	5' AAC CTC T<u>A</u>G GTC CAA 3'

Surface modification with PDC were done as described previously(*4*). Substrates were treated with 0.2% (w/v) PDC (Sigma) in 10% (v/v) pyridine/dimethylformamide (Fisher) for 2 h at room temperature and then washed with HPLC-grade methanol and acetone (Fisher). After drying *in vacuo* at 110°C for 5 min, the PDC-modified glass supports were ready for oligonucleotide attachment.

Attachment of amino-modified oligonucleotides Oligonucleotides to be coupled to glutaraldehyde-treated surfaces were suspended in deionized water, lyophilized in a Speed Vac concentrator (Savant Instruments, Inc., Farmingdale, NY) and resuspended in coupling buffer containing 0.6 g of sodium cyanoborohydride per 100 ml to an oligonucleotide concentration between 1 μM and 50 μM (depending on the experiment). A pattern for oligonucleotide binding regions was created by aligning the modified cover glass on top of a nichrome-patterned glass template, which provided a visual target for pipetting droplets onto the coverslip.

Amino-modified oligonucleotide solutions (0.3 μl) were deposited with a pipette on the modified glass surface at locations defined by the template. This resulted in oligonucleotide spots with an average diameter of < 1 mm. To prevent evaporation, glass supports were sealed with laboratory film in a Petri dish containing water-saturated filter paper for 4 hr . Supports were then rinsed with 10 ml each of coupling buffer, deionized water, 1 M NaCl and deionized water to remove non-covalently bound DNA. Any residual reactive groups on the surface were blocked by reaction with 1 M Tris-HCl (pH 7.5) at room temperature for 1 h. Oligonucleotide arrays were rinsed with 10 ml each of 1 M NaCl and deionized water, air dried at room temperature, and used for hybridization experiments.

Oligonucleotides to be coupled to the PDC surface were lyophilized and resuspended in 100 mM sodium carbonate/bicarbonate buffer (pH 9.0; 1 μM to 50 μM) and were deposited in a pattern as described above. Bound-oligonucleotides were incubated at 37°C for 1 h in a sealed moist Petri dish, then rinsed once with 10 ml of 1% (v/v) NH_4OH and three times with 10 ml deionized water to remove non-covalently bound DNA. Remaining reactive groups on the surface were blocked as previously described and air-dried. Reactions performed in the glass pre-treatment and coupling reactions are summarized in Fig. 1.

Hybridization of small oligonucleotides DNA targets used for hybridization were diluted in 6X SSPE (0.9 M NaCl, 0.06 M Na_2HPO_4, 6 mM EDTA, pH 7.4) containing 1 mM cetyltrimethyl ammonium bromide (CTAB), which was used to enhance hybridization kinetics (6[. Hybridization was performed at 46°C unless noted differently. A commercially available micro-chamber (20 μl Probe-Clip, Grace Bio-Labs, Pontiac, MI) was used for hybridization with a fluorescently-labeled target. Following hybridization, the Probe-Clip was removed and arrays were washed with 2X SSPE containing 0.1% (w/v) SDS for 15 min at 46°C to remove non-hybridized oligonucleotide unless noted differently.

Hybridization of a PCR product To study the influence of steric effects, single-stranded, dye-labeled PCR product was generated. A 219 nt region of the b-globin gene containing exon 1 and the 5' end of IUS 1 was amplified with primers, one of which contained a biotin group (forward primer) and the other a fluorescent group (reverse primer). They were as follows: 5'-biotin-AGT CAG GGC AGA GCC ATC TA-3' and 5'-fluorescein-GCC CAG TTT CTA TTG GTC TCC-3'. Oligonucleotides were purified with an oligonucleotide purification column (ABI). Human genomic DNA (500 ng) was amplified with a Perkin Elmer GeneAmp® PCR System 2400 in a

194

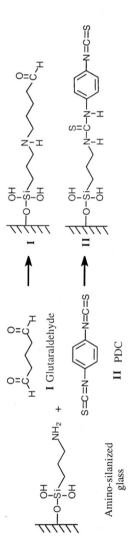

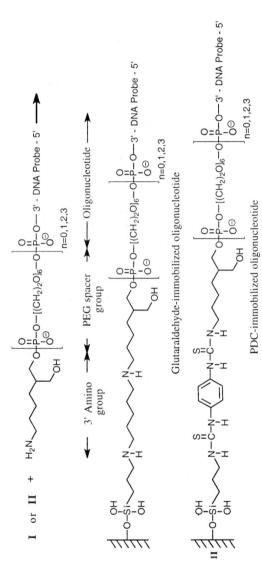

Figure 1. Chemical species, spacers, and reactions involved in coupling nucleotides to glass supports. The two main chemistries used involved gluteraldehyde (I) or PDC (II) as a crosslinking agent.

total volume of 100 ml containing 10 pmol of forward and 25 pmol of reverse primer 200 mM dNTP, 50 mM KCl 10 mM Tris-HCl pH 8.3, 1.5 mM MgCl2 and 0.1% (w/v) gelatin. The cycling protocol was 5 min hot start at 95°C, 30 cycles of 1.5 min at 95°C, 1 min at 58°C and 1.5 min at 72°C, followed by a 5 min final extension at 72°C.

Single-stranded targets were generated from double-stranded PCR product using MPG® Streptavidin beads (CPG, Inc. Lincoln, NJ) as recommended by the manufacturer. Purified PCR products were incubated overnight at room temperature with 40 mL of beads prewashed three times with binding/washing buffer (B/W) (10 mM Tris-HCl, pH 7.5, 1 mM EDTA, 2.0 M NaCl). Beads containing bound PCR products were placed in a magnetic particle concentrator (Dynal MPC, Great Neck, NY). Protocols used for binding the product, washing, and denaturing were those suggested by the manufacturer. This resulted in biotin-labeled strands on the beads and fluorescein-labeled strands in solution.

Image analysis An inverted epifluorescent microscope with a 75 W Xenon lamp (Nikon Diaphot 300, Nikon, NY) equipped with a filter set for fluorescein (XF22, Omega Optical, Brattleboro, VT), a 40X objective (Nikon Fluor 40), and a computer-controlled stage (Nikon TMD/TME Automated Stage) was used for imaging. A CCD camera (TEK 768, Princeton Instruments, Princeton, NJ) was coupled to the side port of the microscope with a Diagnostic Instruments 0.6X lens system. Standard operation of the camera rejects heat to room temperature air and cools the CCD chip to -47°C to decrease dark current and improve the S/N ratio. Lower temperatures (to -67°C) were obtained for some studies by circulating refrigerated coolant through the camera heat exchanger.

For hybridization studies, neutral density filters (Nikon ND2 and ND4) with a combined transmittance of 12.5% were used to decrease the excitation light intensity and to minimize photobleaching. Illumination field was reduced approximately 50% by partially closing field aperture, which reduced significantly stray light. The stage was programmed to sample at 300 mm intervals in a non-overlapping 3 x 3 x-y pattern across each spot for 9 data points per sample. Data acquisition and stage control were automated with a script written in IPLab Spectrum software (Signal Analytics Corp., Vienna, VA), and subsequent image analysis was done on a Power Macintosh 7100/80 computer. All images were obtained with a 20 sec camera exposure time using a 40X objective in the post-hybridization buffers described above. An image of the array was also captured with a 4X objective (Nikon Plan 4) to provide a permanent record of spot uniformity.

For photobleaching studies of surface diffusion and desorption kinetics, a totally internally reflected beam at 488 nm from an argon ion laser was substituted for the xenon lamp. Methodology and theory have been described previously[17]. Various surfaces including those prepared with the aminopropyl-triethoxy silane (APTES), N-methyl aminopropyl trimethoxy silane (APTMS), dimethyldichloro silane (DMS), and trichlorooctadecyl silane (ODS) were tested for interaction with a fluorescently-labeled oligonucleotide (5' CAG TTC CGA CGG GTT AAA CTC 3' with fluorescein attached at the 5' end). In addition, small samples (~0.35 mg) of Corning CPG porous glass particles (pore radius 135 nm and surface area 34 m^2/g) coated with the same silanes were equilibrated with various concentrations of labeled oligonucleotide. The decrease in solution phase concentration was measured and used to calculate equilibrium adsorption isotherms. Quantum 24 Fluorescein Microbead Standards (Flow Cytometry; no. samples > 150) with known surface density were used to convert surface fluorescence of glass slide surfaces to molecules of flourescein/cm^2.

Data Analysis Fluorescent intensities (F_i) from CCD images were evaluated by comparing either
1) average intensities ($\overline{F_{avg}}$) corrected for background (F_{bg}):

$$\overline{F_{avg}} = (\sum_{i=1}^{9} F_i - F_{bg}) / 9; \tag{1}$$

2) average raw signals ($\overline{F_{raw}}$):

$$\overline{F_{raw}} = (\sum_{i=1}^{9} F_i) / 9; \tag{2}$$

or 3) ratios (F_{ratio}) of intensities on a single array, normalized to the highest average intensity region on the same array ($\overline{F_{max}}$) resulting from a perfectly-matched hybrid:

$$F_{ratio} = \frac{\overline{F_i} - \overline{F_{bg}}}{\overline{F_{max}} - \overline{F_{bg}}} \tag{3}$$

Images of fluorescent probes attached to a surface were analyzed according to Eq. 1. Hybridization was analyzed by comparing raw signals using Eq. 2 when a direct comparison was made with the background. The difference in intensity between perfect complementary hybrids and single-base pair mismatches was evaluated by using Eq. 3. When control regions with foreign DNA gave readings less than regions lacking DNA (background), their values were set to zero to avoid negative values. Data for surface adsorption/diffusion/desorption experiments were analyzed as reported elsewhere(*17*).

Results

Oligonucleotide adsorption on and binding to glass Our results show polystyrene had approximately twice and polycarbonate almost four times the fluorescence of glass (with or without surface modification). Thus, glass seemed the most appropriate substrate for our work. Although adenine, guanine, and cytosine contain aromatic amino groups, we found that the addition of an aliphatic amine to the 3' end of each oligonucleotide increased the reactivity to our modified glass 3-5 fold relative to unmodified oligonucleotides. All subsequent work therefore was done with amino-modified probes. Our first study, to determine what surface density of olinucleotides could be obtained, was done with fluorescein/amino-modified probes.

Normally, one assumes that fluorescence is directly proportional to fluorophore concentration, and that measurement of surface fluorescence provides data on the extent of reaction under different conditions. However, our results show a less than proportional signal at high surface concentrations, suggesting quenching due to intermolecular energy transfer. This was confirmed by mixing labeled and unlabeled amino oligonucleotide in various ratios. The two molecules are expected to react equally with the glass surface, so a plot of fluorescence intensity versus labeling ratio (labeled oligo/total oligo) should be linear in the absence of quenching.

Proportionality was demonstrated, however, only for the very lowest probe concentration (Fig. 2). When the amount of bound probe was increased, the fluorescence decreased progressively more than expected from the slope of the low labeling ratio data (dashed line). All data in the figure were taken for gluteraldehyde-treated surfaces after a 4 hr reaction time. Longer times and/or higher oligo

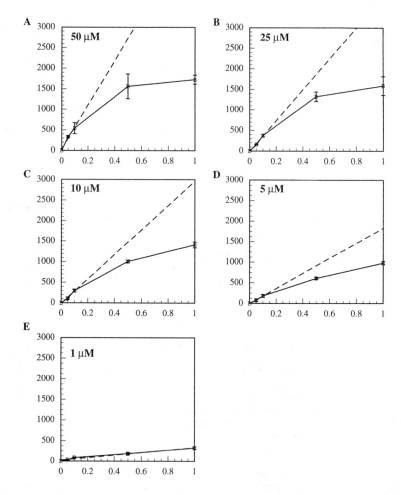

Figure 2. Fluorescent intensities achieved by coupling fluorescent oligonucleotides to glass with gluteraldehyde at different concentrations and with different labeling ratios. Deviation from linearity at higher concentrations provides evidence for quenching. Each data point is the average of 9 images (background subtracted).

concentrations led to greater quenching (data not shown). Our results also show that PDC coupling produced similar effects (Fig. 3). The standard reaction time for this coupling procedure was 1 rather than 4 hrs.

Surface densities were estimated by comparison of unquenched fluorescence readings with the bead standards. For the gluteraldehyde-linked dye, the onset of quenching occurred at about 1.4×10^{11} molecules/cm^2, an inter-dye separation distance of approximately 29 nm. The PDC surfaces exhibited quenching at 3×10^{11} molecules/cm^2, a separation of 20 nm. By extrapolation, we estimated at 50 μM that the density of bound oligonucleotides was 1.3×10^{12}/cm^2 and 1.5×10^{12}/cm^2 for gluteraldehyde- and PDC-linked molecules, respectively. No attempt was made to pack more molecules on the surface with higher concentrations or longer reaction times since this led to severe quenching. From solution-phase data, it is known that fluorescein quenching begins at intermolecular separations of ~8 nm(18). This fiinding suggests that oligonucleotides may have been clumped or clustered on the surface. Our lower limit of detection was ~2×10^{10} molecules/ cm^2.

The physisorption experiments on porous APTES glass gave roughly Langmuirian isotherms. For solution phase concentrations (C) and surface concentrations (A), the data was well represented by $C = K_1 A/(K_2 + A)$ where K_1 was found to be 1.65×10^{13} molecules/cm^2 and K_2 was 16 μM.(17) Our data also strongly suggested that the molecules were lying down flat on the surface. Estimates of the maximum possible monolayer surface coverage for a 21 nt oligonucleotide in this configuration, corrected for mutual interference (the so called "jamming limit"), gave a figure of about 1.46×10^{13} molecules/cm^2. We were able to approach this limit (1.16×10^{13} molecules/cm^2) at a solution-phase concentration of 44.6 μM.

Hybridization of targets to oligonucleotide probe arrays In the application of probe arrays, the apparent kinetics of hybridization are important and dependent on size of diffusing molecules. Among the processes that can influence kinetics are rate of diffusion of molecules to the surface, kinetics of hybridization, and rate of denaturation of mismatched hybrids. Often, intrinsic hybridization kinetics are rate limiting.(19) A short (15 nt) fluorescently-labeled complementary target was studied first. In this case, probe bound to the surface was not labeled. Results for two different hybridization temperatures, 22 and 46 C, and for PDC-coupled probes are shown in Fig. 4. The higher temperature produces more hybridization after 2 hrs, but beyond this time the intensity decreases. Presumably, this is another manifestation of fluorescence quenching as too many labeled molecules crowd together on the surface. Subsequent experiments were performed at 46 C for 2 hours to minimize such effects.

Washing to remove incorrectly-hybridized oligonucleotides also plays an important role in achieving good results. It was found that a 2X SSPE buffer (rather than the 6X SSPE used for hybridization) gave good signals that were 3 times those of background or negative controls. In addition, this solution provided good discrimination (ratio of 1.3 or better) between a perfectly-matched hybrid and one with a single base mis-match. Detergents such as sodium dodecyl sulfate (SDS) and cetyl trimethyl ammonium bromide (CTAB) also are often used to improve performance in similar hybridizations such as Southern blotting. These were tested in our micro-hybridization system, and it was found that 0.1% SDS (w/v) was particularly effective. Although adding detergent decreases the fluorescence signal, it improves discrimination.

200

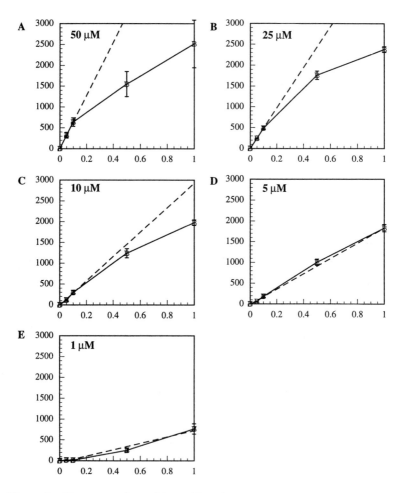

Figure 3. Fluorescent intensities achieved with PDC coupling. This data is otherwise comparable to that in Fig. 2.

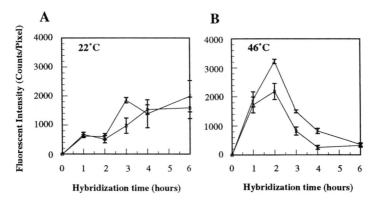

Figure 4. Hybridization between normal human ß globin codon 6 probes and complementary fluorescent targets. The probes were PDC coupled and target was present at a concentration of 2 µM. Data were analyzed according to Eq. 1.

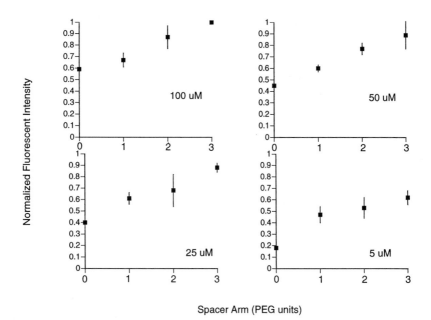

Spacer Arm (PEG units)

Figure 5. Probe spacer arms increase hybridization to a 219 nt target. Normalized fluorescence intensities for a set of three chips in which the number of spacer arm units was systematically varied from zero to three (18 bond lengths per spacer) at various probe concentrations used during the surface linkage step (probe surface densities proportional to this concentration).

Effects of spacer-arm length on hybridization Previous studies show that longer stretches of DNA hybridize more slowly than shorter oligonucleotides, and that addition of a spacer between the probe and surface improves probe accessibility. To study this effect, we had probes synthesized with 0, 1, 2, or 3 spacer units at the 3' end. The spacers were "Spacer Phorphoramidite 18" and were followed by the amino coupling group "3' Amino Modifier C7," (both from Glen Research, Sterling VA). Six sequential ethylene glycol units were present in the spacer, so the "18" signifies that the length of the spacer is approximately 18 carbon single bonds. Likewise, the "C7" indicates that there is a distance between the amino group and the 3'-terminal nucleotide of about 7 bond lengths. Synthesis was carried out by Integrated DNA Technologies, Inc. (Coralville, IA). The structures of these special phosphoramidites are shown in Fig. 1.

Fig. 5 shows that for the four concentrations used when linking probes to the surface, hybridization with the 219 nt target increased monotonically with spacer length, and that over the spacer size range studied, there was an approximate doubling of fluorescent signal intensity from no spacer to the 3 spacer unit probe.

Discrimination between perfect matches and single-base mismatches Although our goals were primarily to obtain fundamental information about the conditions necessary for oligonucleotide binding and hybridization, we felt that it was also important to demonstrate under optimized conditions that we could discriminate between single-base mismatches and perfect probe/target matches. Identical chips were made in which the first row (A) contained an oligonucleotide from the ß globin gene which contained the normal sequence at codon 6, the second (B) a codon 6 mutant (A/T), the third (C) codon 39 normal, and the last (D) a codon 39 mutant (C/T). Table I gives the entire sequences. These were contacted with solutions of the four complementary sequence targets that were fluorescently labeled.

Good discrimination between perfect matches and single-base mismatches was possible, while negative controls gave very little signal under identical conditions. Signal intensity averages and standard deviations from 8 arrays using glutaraldehyde coupling for perfect matches, single-base pair mismatches and negative controls were 0.88± 0.17, 0.42±0.22 and 0.01±0.07, respectively. Averages and standard deviations from 8 arrays using PDC coupling for the same sequence were 0.85±0.16, 0.31±0.16 and 0.039±0.14. These ranges were non-overlapping, and the averaged raw signal ratios (perfect match to mismatch) for glutaraldehyde and PDC were 1.6 and 1.8, respectively. These results show that under the proper conditions, arrays can discriminate between single-base mismatches and perfect probe/target matches.

Discussion

Covalent attachment of oligonucleotides to glass Although both glutaraldehyde and PDC methods produced high oligonucleotide surface coverage on amino-silanized glass, in our hands the PDC method had several advantages. It was four times faster for the same level of surface coverage and was easier to use. In addition, PDC-activated glass could be stored for two months and still coupled amino-modified oligos quite well. Finally, the PDC-coupled probes were slightly better at discriminating perfect matches from single-base mismatches than those prepared with glutaraldehyde. Both chemistries could be used to cover a 50-fold range of probe concentrations, although quenching limited the useful measurement range to approximately 10-fold. Mixing unlabeled probes with those that have been labeled can facilitate identification of quenching effects. Fluorescein also exhibited troublesome photo-bleaching properties, which affected the lower limit of detection that we could achieve with this dye.

One group has reported achieving 6 to 10 times more surface coverage (9.8×10^{12} to 1.4×10^{13} molecules/cm^2) of DNA than we observed using a hetero-bifunctional crosslinker ($20,21$). Our studies of maximal physisorption on porous glass gave results similar to the covalently-bonded oligonucleotide surface coverage reported by these authors. However, it is not clear that one should aim for such high surface coverages. Others recommend partial removal of excess oligonucleotides on a surface, and state that too much oligonucleotide interferes with hybridization (22). The methods we employed, measurement of fluorescently-labeled oligonucleotides by solution depletion over porous particles and microscopic examination of surfaces with a sensitive camera, should prove useful as others try to optimize surface probe densities.

Effect of spacer-arm length Our studies on effects of 0, 18, 36, and 54 effective carbon bond spacers demonstrated that they provided improved hybrid formation between probe and target. A recent study confirms this general trend(22). However, in contrast to the approximate doubling of the hybridization signal that we observed, this group found that the signal for probes with 40 atom spacers was enhanced as much as 150-fold over no spacer (for sparsely covered surfaces). They also reported an optimum spacer length. Differences in surface chemistry, surface density, target and solution conditions may account for the differences between our findings and theirs. All of these results point to the importance of obtaining still more surface chemical information on oligonucleotide probes and hybrids they form.

Hybridization and mismatch discrimination The optimized hybridization conditions we found provided for detection of single-base mismatches. Our results were comparable to discrimination levels reported by others. In our studies, the mismatch was located near the center of a probe sequence, which maximizes its detection(5). In this case, the mismatches represented mutations of codons 6 and 39 in the ß-globin gene, representing a clinically relevant application.

Conclusions

Although oligonucleotide or cDNA probes are now being used almost routinely for DNA and RNA analysis by hybridization, much remains unknown about some of the basic physical and chemical events occurring on these arrays. Our studies and those of others are beginning to show the importance of surface probe densities, steric effects related to spacer and attachment chemistry, and other factors such as the detection scheme employed. Fundamental work is likely to have a strong influence on the eventual success of this important technology.

Acknowledgment

This work was supported by a Whitaker Foundation special opportunities grant, the University of Pennsylvania Research Foundation and Cancer Center Pilot Projects Program, The Foerderer Fund for Excellence from the Children's Hospital of Philadelphia, NIH grants R01 DK 16691 and P60-HL38632, the Nemours Foundation, and a fellowship from the American Association of University Women. We thank Drs. E. Rappaport and H. Kazumi for their assistance with oligonucleotide preparation and microscopic image analysis, respectively.

Literature Cited

1. Shalon, D.; Smith, S. J.; Brown, P. O. *Genome Research* **1996**, *6*, 639-645.
2. Schena, M.; Shalon, D.; Davis, R. W.; Brown, P. O. *Science* **1995**, *270*, 467-470.
3. Maskos, U.; Southern, E. M. *Nucleic Acids Research* **1993**, *21*, 2269-2270.
4. Guo, Z.; Guilfoyle, R. A.; Thiel, A. J.; Wang, R.; Smith, L. M. *Nucleic Acids Research* **1994**, *22*, 5456-5465.
5. Lipshutz, R. J.; Morris, D.; Chee, M.; Hubbell, E.; Kozal, M. J.; Shah, N.; Shen, N.; Yang, R.; Fodor, S. P. A. *BioTechniques* **1995**, *19*, 442-447.
6. Cronin, M. T.; Fucini, R. V.; Kim, S. M.; Masino, R. S.; Wespi, R. M.; Miyada, C. G. *Human Mutation* **1996**, *7*, 244-255.
7. Kozal, M. J.; Shah, N.; Shen, N.; Yang, R.; Fucini, R.; Merigan, T. C.; Richman, D. D.; Morris, D.; Hubbell, E.; Chee, M.; Gingeras, T. R. *Nature Medicine* **1996**, *2*, 753-759.
8. Sapolsky, R. J.; Lipshutz, R. J. *Genomics* **1996**, *33*, 445-456.
9. Saiki, R. K.; Chang, C.-A.; Levenson, C. H.; Warren, T. C.; Boehm, C. D.; Kazazian, H. H.; Ehrlich, H. A. *New England Journal of Medicine* **1988**, *319*, 537-541.
10. Kern, W.; Puotinen, D. A. *RCA Review* **1970**, *June*, 187-206.
11. Lamture, J. B.; Beattie, K. L.; Burke, B. E.; Eggers, M. D.; Ehrlich, D. J.; Fowler, R.; Hollis, M. A.; Kosicki, B. B.; Reich, R. K.; Smith, S. R.; Varma, R. S.; Hogan, M. E. *Nucleic Acids Research* **1994**, *22*, 2121-2125.
12. Fodor, S. P. A.; Read, J. L.; Pirrung, M. C.; Stryer, L.; Lu, A. T.; Solas, D. *Science* **1991**, *251*, 767-73.
13. Chrisey, L. A.; Roberts, P. M.; Benezra, V. I.; Dressick, W. J.; Dulcey, C. S.; Calvert, J. M. *Materials Research Society Symposium Proceedings* **1994**, *330*, 179-184.
14. Williams, R. A.; Blanch, H. W. *Biosensors and Bioelectronics* **1994**, *9*, 159-167.
15. Hermanson, G. T.; Mallia, A. K.; Smith, P. K. *Immobilized Affinity Ligand Techniques*; Academic Press, Inc.: San Diego, 1992.
16. Pontius, B. W.; Berg, P. *Proceedings of the National Academy of Sciences USA* **1991**, *88*, 8237-8241.
17. Chan, V.; Graves, D. J.; Fortina, P.; McKenzie, S. E. *Langmuir* **1997**, *13*, 320-329.
18. Kalb, E.; Frey, S.; Tamm, L. K. *Biochim. Biophys. Acta* **1992**, *1103*, 307.
19. Chan, V.; Graves, D. J.; McKenzie, S. E. *Biophysical Journal* **1995**, *69*, 2243-2255.
20. Chrisey, L. A.; O'Ferrell, C. E.; Spargo, B. J.; Dulcey, C. S.; Calvert, J. M. *Nucleic Acids Research* **1996**, *24*, 3040-3047.
21. Chrisey, L. A.; Lee, G. U.; O'Ferrall, C. E. *Nucleic Acids Research* **1996**, *24*, 3031-3039.
22. Shchepinov, M. S.; Case-Green, S. C.; Southern, E. M. *Nucleic Acids Research* **1997**, *25*, 1155-1161.

Chapter 15

Application of Neural Networks to Optimize Flux Profiles for Transdermal Systems

R. T. Kurnik, J. Jona, and B. Jasti

Cygnus, Inc., 400 Penobscot Drive, Redwood City, CA 94063

Neural networks are being increasingly used in the pharmaceutical field for many different applications due to their ability to model processes that cannot be modeled by classical methods. A new area for application of neural network technology is in the design and optimization of controlled release systems, either oral, implant, or transdermal. In these systems, a common requirement is to determine a formulation that matches a pre-determined profile, typically release into water, or transport through skin, consistent with constraints of cost, skin adhesion, skin irritation, and adhesive cold flow. In this paper, we apply neural networks to determine the chemical composition of an estradiol transdermal device such that the resultant skin transport profile matches a desired profile.

In controlled drug delivery, it is desirable to obtain a specific delivery profile for the drug of interest. Typically, this is performed by a combination of experimental design methodology and trial and error. Although this approach is satisfactory, it is very time consuming and allows one to get a reasonable, but not necessarily optimal formulation. Consequently, there is great interest in finding more efficient methodologies to determine formulations and to have these formulations work right the first time. One approach that is useful in this regard is to use neural network technology in the analysis phase of an experimental design. In the current work, a neural network is used to optimize the formulation of an estradiol transdermal delivery device.

Achieving the targeted profile for the delivery of a transdermal drug is the first step in the development of a therapeutic system. The other important factors to be considered are the physical properties and the stability of the transdermal system. The physical properties of our system are determined by the drug, enhancer, crystallization inhibitors, and fillers. In general, an enhancer will increase the flux of drug across the stratum corneum. However, the enhancer may also plastisize the adhesive and lead to cold flow, thus requiring a delicate balance on the amount of enhancer.

Increasing the percentage loading of the drug (in this case estradiol) will result in a higher delivery up to a certain loading after which time the delivery will plateau. The loading of the drug is limited by its solubility in the adhesive and also by economic factors. On the other hand, too high a loading may lead to crystallization either immediately, or upon storage and thus lead to a change in the delivery profile. One can increase the loading of the drug by incorporating crystallization inhibitors, such as PVP (*1*), however, too much PVP will decrease the skin flux due to the increased solubility of estradiol in PVP. Due to these many interactions between the formulation variables, including the need to match a predetermined skin transport profile, it is desirable to have a structured approach, such as a neural net methodology, to guide the formulator in designing an optimal product.

Neural Network Design and Training

Neural networks are used to determine interrelationships between variables based on a pattern recognition method that attempts to mimic that of the human brain (*2-7*). The neural network is composed of many interconnected processing elements, each of which is a mathematical analogue of a neuron. A typical network is composed of one input layer, one or more hidden layers, and one output layer.

The development of a neural network has two aspects: A training mode and an application mode. In the training mode, the input variables are presented to the neural network and predictions of the output are made. The training is continued until some minimization criteria for the error of the predicted output compared to the experimental output is realized. Once a trained network is established, the input variables can be varied and the value of the output variable established. Generally, the neural network predictions are only valid for input variables within the range used for training.

Many different architectures exist for neural networks, and backpropagation networks are most commonly employed (*8*), as was the case in this study. A simplified backpropagation network is illustrated in Figure 1, where in this case there is one input layer consisting of four composition variables plus time, one hidden layer, and one output layer, which in this case is the transport across skin. This three layer approach is able to determine the (nonlinear) patterns that map the input signals to the output result by determining appropriate weighting factors at each of the input and hidden layer nodes.

The mathematical procedure used to determine the input to hidden layer and then the output layer mapping is as follows (*9*): Briefly, a neural element, as shown in Figure 2, receives input from the previous layers or directly from the input data. A weighted linear combination of the inputs is formed and typically a logistic function is applied to the weighted sum of inputs. The logistic function is described by equation 1

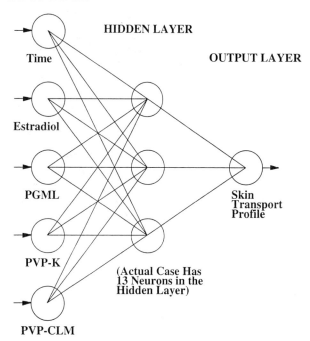

Figure 1. Neural net architecture for backpropagation.

208

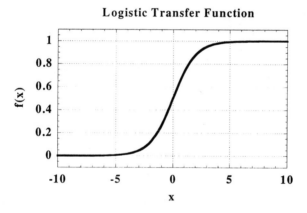

Figure 2. Typical neural element and transfer function.

$$f(x) = \frac{1}{1+e^{-x}}.$$ (1)

During the training phase, the weights are adjusted so as to minimize the sum of the squares deviation between the experimental and neural network prediction. This procedure is as follows: Inputs X_i^{s-1} from the previous layer $(s-1)$ are combined by a linear transformation as shown in equation 2

$$I_j^s = \sum_i w_{ji}^s X_i^{s-1}$$ (2)

to give the initial output I_j^s for the given node on the current layer (s). In equation 2, w_{ji}^s is the weight connecting the ith element in the previous layer and the jth element in the current layer. The final output for the current node, x_j^s, is obtained by applying the non-linear transfer function to I_j^s according to equation 1 to give the result

$$x_j^s = f(I_j^s)$$ (3)

The unique feature of neural nets is their capability to self-learn as compared to traditional methods, such as non linear regression. Self learning is accomplished by adjusting each weight such that the resulting output mimics the desired output as closely as possible by using an appropriate learning rule. The learning rule that is typically used is the delta rule which is defined as

$$e_j^s = f'(I_j^s) \sum_k e_k^{s+1} w_{kj}^{s+1}$$ (4)

where
e_j^s is the error for the jth element in the sth layer, $f'(I_j^s)$ is the first derivative of the transfer function, and w_{kj}^{s+1} is the connecting weight between the jth element in the sth layer and the kth element in the $s+1^{th}$ layer.

The weights are generally updated according to the rule

$$w_{ji}^{s,n+1} = \eta e_j^s x_i^{s-1} + \alpha(w_{ji}^{s,n} - w_{ji}^{s,n-1})$$ (5)

where η is the learning rate coefficient, α is the momentum factor, and n is the iteration number. The learning rate influences how much the weight factors are modified the next time the same pattern is presented. For a learning rate of 0.1 (which was used in this study) the weight change would be one-tenth the error.

Caution must be used, however, when increasing the learning rate, as too large a rate leads to oscillation of weight changes. One way to allow faster learning without oscillation is to make the weight change a function of the previous weight change, which provides a smoothing effect. The momentum factor (which was 0.1 in this study) determines the proportion of the last weight change that is added into the new weight change.

In summary, neural networks use linear and non-linear transformations to form a network that attempts to mimic the human neural system. The number of layers and number of nodes in each layer are system dependent. Input data moves from the input layer to the hidden layer(s) and then to the output layer, where the resulting output is compared to the experimental output. The error is then propagated back through the network and is used to update the weights.

Materials and Methods

Materials. Acrylate adhesive was obtained from National Starch and Chemical Company (Bridgewater, NJ). Estradiol was obtained from Sigma Chemical Company (St. Louis, MO). Propylene glycol monolaurate (PGML) was obtained from Gattefosse (San Preist, France). Polyvinylpyrrolidone (PVP-K) was obtained from BASF (Parsippary, NJ). Cross-linked-micronized polyvinylpyrrolidone (PVP-CLM) was obtained from ISP technologies, Inc (Wayne, NJ). All the chemicals were used as received. Human cadaver skin was obtained from a hospital morgue with signed consent and tested negative for HIV and hepatitis B. Experimental design software was from S-Matrix (S-Matrix Corporation, Cupertino, CA) and neural network software was from Ward Systems (Frederick, MD).

Experimental design. An experimental design (S Matrix mixture model with Scheffe criteria and resolution IV, with one boundary constraint) was generated with four formulation components consisting of estradiol (E2) loading from 1.5-3%, propylene glycol monolaurate (PGML) from 10-20%, polyvinylpyrrolidone (PVP-K) from 0-20%, cross-linked-micronized polyvinylpyrrolidone (PVP-CLM) from 0-20%, and the remainder of the composition consisting of the adhesive material. A boundary constraint of

$$5 \leq \%(PVP\text{-}K) + \%(PVP\text{-}CLM) \leq 15$$

was also imposed. This boundary constraint was imposed because when PVP is present below 5%, estradiol forms crystals and when present above 15%, patch adhesion to skin drops significantly. The experimental design consisted of fourteen formulations.

These formulations were made in the laboratory and subsequently tested for flux of estradiol through the skin. The role of PGML in these formulations is that of a permeation enhancer for the transdermal delivery of estradiol as well as other drugs (*10*), while PVP-K and PVP-CLM are fillers that effect the physical properties (adhesion) and also play a role in preventing crystallization of estradiol upon storage.

Transdermal patch preparation. Estradiol was dissolved in an ethanol:ethyl acetate mixture. PGML was added and then the PVP-K and/or the PVP-CLM was added. In the case of PVP-K, the solution was sonicated for a few minutes until a clear solution was obtained; for PVP-CLM, the mixture was sonicated for about 30 minutes at room temperature. The acrylate adhesive was then added and mixed well by hand using a wooden spatula and left rotating overnight. After the overnight mixing for both adhesives, the solution was cast with Gardner Knife on a release liner and dried in the oven for 40 minutes at 70°C. It was then cooled at room temperature and the backing was laminated to the top.

Skin flux studies. These were conducted using human cadaver skin. The cells were modified Franz diffusion cells and the receiver fluid was 0.9% NaCl with 0.01% NaN_3 to prevent microbial growth. At each time point over the seven-day period, the entire content of the receiver was sampled and replaced by fresh solution to maintain diffusional sink conditions. The patch size used for all the studies was 3/8 inch (~0.9 cm²) over the skin diffusion area of 0.71 cm² to avoid edge effects. The samples were analyzed by an HPLC. Each formulation was run in triplicate.

Assay procedure. The analysis of the permeation experiment was performed by a HPLC system. The HPLC separations were conducted using a programmed pump (model LC620), an autoinjector (model ISS 100C) and an ultraviolet-visible diode array (model LC-235) operated at 230nm wavelength, all from PerkinElmer, USA. The separations were accomplished on a Novapak C-18 column (150 mm X 3.9 mm) using isocratic mode. The mobile phase consisted of 55% water and 45% acetonitrile at a flow rate of 1.0 ml/min. The retention time was 3.2 minutes.

Neural Network Analysis

A single hidden layer backpropagation neural network as illustrated in Figure 1 was used to fit the experimental data. This method had 5 variables in the input layer, 13 variables in the hidden layer, and one output variable. There were 126 sets of input data used to train the neural network. A judicious choice of the number of hidden neurons must be made. If too many neurons are used, the system will mimic each data point, but have little predictive capability; if too few hidden neurons are used, the system will be unable to characterize the relationships between variables.

The equation relating the number of hidden neurons to the number of data points, input neurons, output neurons, and the degree of over determination is given by (9)

$$J = \frac{\dfrac{N}{\alpha} - K}{I + K + 1} \qquad (6)$$

where N is the number of data sets, I is the number of input neurons, J is the number of hidden neurons, K is the number of output neurons, and α is the degree of overdetermination ($\alpha = 1.3$ corresponds to a 30 % overdetermined system). The term overdetermined means that there are more data points than unknown parameters. Typically, the degree of overdetermination should be between 20% to 100% to enable a reasonable predictive capability of the neural network. The neural network software used (NeuroShell 2) picks a default number of neurons according to the equation

$$ J = \frac{1}{2}(I + K) + \sqrt{N} \tag{7} $$

The software selected 15 as the default number of neurons, whereas using equation 6 and a 40% degree of overdetermination, the number of hidden neurons selected is 13, which was the number used in this study.

As the logistic transfer function, equation 1, maps to the range (0-1), all input and output variables (including time) need to be normalized to the range (0-1) prior to training. This is accomplished by subtracting from each variable the smallest value in the group and dividing by the range of the group (maximum - minimum). Evaluation of the trained network is then accomplished by rescaling the input - output pairs.

Training of the neural net was done using Neural Shell2, release 3.0 from Ward Systems group (Frederick, MD) and was trained until the average error was less than 0.01%. The correlation coefficient of the actual data vs. the neural net prediction was $r^2 = 0.996$, with a slope of 0.996 ± 0.008, and a y-intercept of 0.002 ± 0.003. A scatter graph of this data is shown in Figure 3, which shows a good fit to the training data set.

Results and Discussion

Shown in Figure 4 is one (of the fourteen) input transport profile sets showing the cumulative estradiol skin transport vs. time. The transport profiles consisted of 9 data points over time. The data were rescaled using the same normalization factors as for the input data. As can be seen, the agreement between the observed data and neural net prediction is excellent and is typical of all fourteen input data sets.

Figure 5 shows the desired skin transport profile as well as a set of conditions (and resultant profile) as determined from the neural network model that resulted in a match to the desired profile. These conditions were determined by implementing the trained net in an Excel spreadsheet (by use of appropriate Microsoft *.dll files generated with the Neural Shell2 software). By running the trained network within a spreadsheet, it is very easy to overlay predictions on top of the desired profile. The neural network was also able to determine two additional formulations that resulted in a match to the desired profile.

It should be noted that the desired profile did not overlap any of the observed profiles obtained from the experimental design conditions, so that identification of the optimum formulation parameters was not obvious. All of the

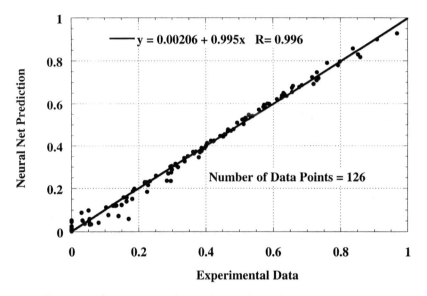

Figure 3. Scatter plot of experimental data vs. neural network prediction.

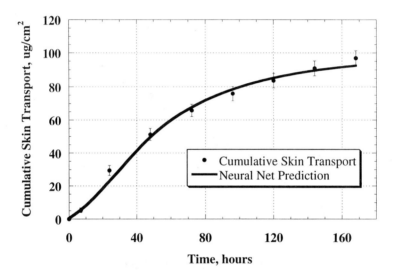

Figure 4. Comparison of experimental data for cumulative skin transport and a neural network prediction for this data.

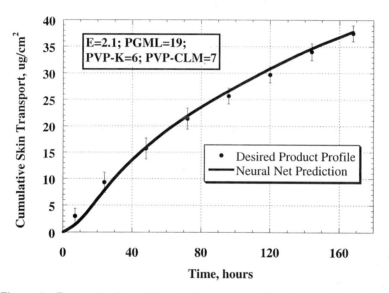

Figure 5. Determination of formulation composition that results in the desired cumulative skin transport profile for a full size transdermal patch.

optimized input parameter settings that were used to fit the desired profile were within the range used for training the neural net and would therefore be expected to give an accurate prediction.

In Figure 6 are shown the desired skin transport profile and a formulation from the neural network that matches this profile, but using one-half the transdermal patch size, i.e. the predicted profile is divided by 2. This is a significant finding, as a smaller patch size is more desirable to the patient and is more economical. Four different formulations were found that would meet these conditions, as shown in Table I. These formulations are intuitively sensible. For example, as the estradiol loading is decreased at a constant level of PVP-CLM, the PGML concentration must increase to maintain the same transport profile. Also, at a constant estradiol loading, as the percentage PGML is increased, PVP-K also increases due to the partitioning of PGML into PVP-K.

Table I. Formulation Compositions for a One-Half Size Patch

	Composition
Formulation I	E=2.5; PGML=14; PVP-K=11; PVP-CLM=3
Formulation II	E=2.1; PGML=19; PVP-K=12; PVP-CLM=3
Formulation III	E=2.0; PGML=20; PVP-K= 9; PVP-CLM=3
Formulation IV	E=2.0; PGML=15; PVP-K= 4; PVP-CLM=3

Choice of the formulation from the four formulations presented in Table I depend upon the physical properties and stability of the final system. For example, formulation I containing 2.5% estradiol and 11% PVP-K may have less skin adhesion than formulation III containing 2.0% estradiol and 9% PVP-K. In choosing between formulation III and IV, which have the same estradiol loading but different PGML and PVP-K loadings, formulation III would typically be chosen as the higher PVP-K content would prevent crystallization of the estradiol.

Conclusions

Neural networks provide an alternative approach for empirical model building for data. In this application, neural networks were shown to be useful in the design and optimization of a controlled release system. The key feature of neural networks is their ability to learn non-linear variable interactions so that the effect of all components in a formulation on a transport profile can be ascertained. The best way to obtain data to train a neural network is to use input - output pairs from an appropriately characterized experimental design. The trained network should then be able to determine multiple formulations such that the resultant transport profile will overlap a predetermined profile.

In the example discussed in this paper, multiple formulations (for both the full size patch and a one-half size patch) were found that would yield the required

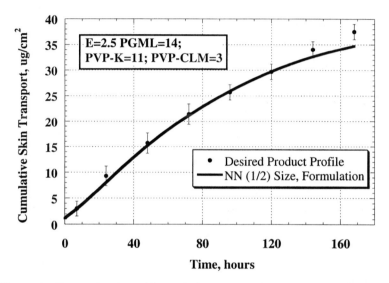

Figure 6. Determination of formulation composition that results in the desired cumulative skin transport profile for a one-half size transdermal patch.

skin transport profile. The final choice of a formulation would depend on other factors, such as cost, skin adhesion, skin irritation, and adhesive cold flow. Having at hand several formulations to choose from, gives the formulation chemist a significant advantage to determine the optimum formulation.

Acknowledgments

The assistance of Bret Berner, Chia-Ming Chiang, and Russell Potts in the preparation of this manuscript is greatly appreciated.

Literature Cited

1.   X. Ma, Taw, J., Chiang, C.-M., *Int. J. Pharm.,* **1996**, *142*, 115.
2.   R. J. Erb, *Pharmaceutical Research,* **1993**, *10*, 165-170.
3.   A. S. Hussain, Yu, X, Johnson, R.D., *Pharmaceutical Research,* **1991**, *8*, 1248-1252.
4.   *Parallel Distributed Processing: Explorations in the Microstructure of Cognition. Vol 1. Foundations;* D. E. Rumelhart, McClelland, J.L., Eds.; MIT Press: Cambridge, MA, 1986.
5.   T. Aoyama, Ichikawa, H., *Chem-Pharm-Bull, (Tokyo),* **1991**, *39*, 358-366.
6.   S. P. Chitra, *AI Expert,* **1992**, *7*, 20-25.
7.   J. A. Freeman, *AI Expert,* **1992**, *7*, 27-35.
8.   J. A. Freeman, *Simulating Neural Networks with Mathematica;* Addison-Wesley: Reading, MA, 1994.
9.   W. C. Carpenter, Hoffman, M.E., *AI Expert,* **1995**, *10*, 30-33.
10.   G. W. Cleary, Roy, S., *United States Patent,* Cygnus Research Corporation, U.S.A., 1990.

Author Index

Subject Index

A

Accelerated release
examples of guest release by
cyclodextrins, 124, 125t
See also Cyclodextrin technology
Acetaminophen
gel and solute ionic character at high and
low pH, 44t
loading in nonionic poly(ethylene oxide)
gel-coated catheter, 48
See also Gel-coated catheters
Acid-catalyzed triggering
folate-targeting to KB cells and pH-
induced endosomal release, 174, 177
pH-dependent hydrolysis kinetics of
diplasmenylcholine and relationship to
liposomal release rates, 171, 174
plasmenyl-type liposomes, 171, 174, 177
See also Plasmenyl-type liposomes
Acrylamide (AAm)
copolymer in polymer hydrogels, 15, 17
See also Poly(*N,N*-dimethylaminoethyl
methacrylate-*co*-acrylamide)
(polyDMAEMA-*co*-AAm)
Adsorption of DNA. *See* DNA adsorption
and hybridization on solid surfaces
Aggregate transport
addition of solution or suspension on skin,
80
barrier resistance, 80, 82
change in hydration energy, 83
deformability, 83–84
equations describing transport across non-
occluded skin, 79, 82
external and trans-barrier gradients, 79
hydration gradient, 82–83
local elastic energy of transfersome
membrane, 83–84
schematic representation of forces
affecting material transport, 81*f*
stress-dependent rigidity factor, 82

suspension-driving pressure, 84
water flux, 79–80
See also Artificial smart drug carriers
Angina, chronopharmacology, 87
Angioplasty catheters. *See* Gel-coated
catheters
Artificial smart drug carriers
aggregate transport, 79–84
aggregate transport from transfersome
across artificial skin barrier, 73*f*
average glucose concentration change in
blood as function of time after
administration of insulin, 78*f*
barrier resistance to aggregate transport,
80, 82
deformability of carriers, 83–84
difficulty of skin crossing, 70–71
effect of water activity difference across
barrier on trans-barrier flow of marker,
74*f*
efficacy of tritiurated dipalmitoyl
phosphatidylcholine transfer across
intact murine skin in vivo, 73*f*
hydration gradient, 82–83
hydrophilicity of typical liposome surface,
72
influence of relative size of liposomes or
transfersomes on vesicle sieving, 74*f*
insulin transport, 77
overcoming size exclusion principle, 85
peptide and protein transport in vivo, 75,
77
schematic representation of forces
affecting material transport, 81*f*
seeking materials for improved delivery,
70–71
skin administration in transfersomes, 85
skin model, 72, 75
skin penetration pathways, 71
small agent transport in vivo, 75, 76*f*
small aggregate formation resulting in
micellization, 72

Bestsellers from ACS Books

The ACS Style Guide: A Manual for Authors and Editors (2nd Edition)
Edited by Janet S. Dodd
470 pp; clothbound ISBN 0–8412–3461–2; paperback ISBN 0–8412–3462–0

Writing the Laboratory Notebook
By Howard M. Kanare
145 pp; clothbound ISBN 0–8412–0906–5; paperback ISBN 0–8412–0933–2

Career Transitions for Chemists
By Dorothy P. Rodmann, Donald D. Bly, Frederick H. Owens, and Anne-Claire Anderson
240 pp; clothbound ISBN 0–8412–3052–8; paperback ISBN 0–8412–3038–2

Chemical Activities (student and teacher editions)
By Christie L. Borgford and Lee R. Summerlin
330 pp; spiralbound ISBN 0–8412–1417–4; teacher edition, ISBN 0–8412–1416–6

Chemical Demonstrations: A Sourcebook for Teachers, Volumes 1 and 2, Second Edition
Volume 1 by Lee R. Summerlin and James L. Ealy, Jr.
198 pp; spiralbound ISBN 0–8412–1481–6
Volume 2 by Lee R. Summerlin, Christie L. Borgford, and Julie B. Ealy
234 pp; spiralbound ISBN 0–8412–1535–9

The Internet: A Guide for Chemists
Edited by Steven M. Bachrach
360 pp; clothbound ISBN 0–8412–3223–7; paperback ISBN 0–8412–3224–5

Laboratory Waste Management: A Guidebook
ACS Task Force on Laboratory Waste Management
250 pp; clothbound ISBN 0–8412–2735–7; paperback ISBN 0–8412–2849–3

Reagent Chemicals, Eighth Edition
700 pp; clothbound ISBN 0–8412–2502–8

Good Laboratory Practice Standards: Applications for Field and Laboratory Studies
Edited by Willa Y. Garner, Maureen S. Barge, and James P. Ussary
571 pp; clothbound ISBN 0–8412–2192–8

For further information contact:
Order Department
Oxford University Press
2001 Evans Road
Cary, NC 27513
Phone: 1-800-445-9714 or 919-677-0977

Highlights from ACS Books

Desk Reference of Functional Polymers: Syntheses and Applications
Reza Arshady, Editor
832 pages, clothbound, ISBN 0–8412–3469–8

Chemical Engineering for Chemists
Richard G. Griskey
352 pages, clothbound, ISBN 0–8412–2215–0

Controlled Drug Delivery: Challenges and Strategies
Kinam Park, Editor
720 pages, clothbound, ISBN 0–8412–3470–1

Chemistry Today and Tomorrow: The Central, Useful, and Creative Science
Ronald Breslow
144 pages, paperbound, ISBN 0–8412–3460–4

Eilhard Mitscherlich: Prince of Prussian Chemistry
Hans-Werner Schutt
Co-published with the Chemical Heritage Foundation
256 pages, clothbound, ISBN 0–8412–3345–4

Chiral Separations: Applications and Technology
Satinder Ahuja, Editor
368 pages, clothbound, ISBN 0–8412–3407–8

Molecular Diversity and Combinatorial Chemistry: Libraries and Drug Discovery
Irwin M. Chaiken and Kim D. Janda, Editors
336 pages, clothbound, ISBN 0–8412–3450–7

A Lifetime of Synergy with Theory and Experiment
Andrew Streitwieser, Jr.
320 pages, clothbound, ISBN 0–8412–1836–6

Chemical Research Faculties, An International Directory
1,300 pages, clothbound, ISBN 0–8412–3301–2

For further information contact:
Order Department
Oxford University Press
2001 Evans Road
Cary, NC 27513
Phone: 1-800-445-9714 or 919-677-0977
Fax: 919-677-1303